Expert GeoServer

Build and secure advanced interfaces and interactive maps

Ben Mearns

BIRMINGHAM - MUMBAI

Expert GeoServer

Commissioning Editor: Richa Tripathi
Acquisition Editor: Denim Pinto
Content Development Editor: Nikhil Borkar
Technical Editor: Subhalaxmi Nadar
Copy Editor: Safis Editing
Project Coordinator: Ulhas Kambali
Proofreader: Safis Editing
Indexer: Aishwarya Gangawane
Graphics: Tom Scaria
Production Coordinator: Arvindkumar Gupta

First published: July 2018

Production reference: 1300718

Published by Packt Publishing Ltd.
Livery Place
35 Livery Street
Birmingham
B3 2PB, UK.

ISBN 978-1-78953-860-1

www.packtpub.com

`mapt.io`

Mapt is an online digital library that gives you full access to over 5,000 books and videos, as well as industry leading tools to help you plan your personal development and advance your career. For more information, please visit our website.

Why subscribe?

- Spend less time learning and more time coding with practical eBooks and Videos from over 4,000 industry professionals

- Improve your learning with Skill Plans built especially for you

- Get a free eBook or video every month

- Mapt is fully searchable

- Copy and paste, print, and bookmark content

PacktPub.com

Did you know that Packt offers eBook versions of every book published, with PDF and ePub files available? You can upgrade to the eBook version at `www.PacktPub.com` and as a print book customer, you are entitled to a discount on the eBook copy. Get in touch with us at `service@packtpub.com` for more details.

At `www.PacktPub.com`, you can also read a collection of free technical articles, sign up for a range of free newsletters, and receive exclusive discounts and offers on Packt books and eBooks.

Contributors

About the author

Ben Mearns was previously the Lead Geospatial Information Consultant at the University of Delaware, where he was the senior GIS professional charged with leading, developing, and advising on GIS solutions. He developed the GeoServer-based campus mapping system. He has held GIS and data positions at the University of Pennsylvania Cartographic Modeling Lab, Princeton University Department of Sociology, and Macalester College Department of Geography.

Packt is searching for authors like you

If you're interested in becoming an author for Packt, please visit `authors.packtpub.com` and apply today. We have worked with thousands of developers and tech professionals, just like you, to help them share their insight with the global tech community. You can make a general application, apply for a specific hot topic that we are recruiting an author for, or submit your own idea.

Table of Contents

Preface 1

Chapter 1: Developing a Spatial Analysis Platform with WPS 5
   Installing and learning the basics of WPS 5
   WPS request builder 11
   Process chaining 14
   OpenLayers integration 21
   Summary 25

Chapter 2: Speed Up Your App with Tile Caching 27
   Tile caching basics 27
   Configuring tile caching in GeoServer 31
   Creating a tile-backed OL app 37
   Using the tile cache 43
   Resolving problems with tile caching 50
   Summary 53

Chapter 3: Optimizing GeoServer 55
   Resolving bottlenecks 55
   Optimizing vector data stores 61
     Vector data formats 64
   Optimizing raster data stores 66
     WCS 72
   Clustered deployment 74
   Summary 78

Chapter 4: Secure Authentication 79
   Configuring the proxy 79
   HTTPS with TLS and certificates 83
   GeoServer authentication 92
   Secure login from OL 97
   Summary 102

Chapter 5: Putting it into Production 103
   Hosting your GeoServer instance and app 103
   Monitoring the GeoServer instance 107
   Backup and recovery 111
   Production checklist 115
     References and credentials 116
     Locking down 116

Regular maintenance 117
Tuning 117
Summary 118

Other Books You May Enjoy 119

Index 123

Preface

This book will cover web-processing services, speeding up your app with tile caching, optimizing GeoServer to get the best and most stable performance, and putting your GeoServer-backed app into production.

By the end of the book, you will understand how to put a GeoServer-backed app into production on a host infrastructure with the best performance and reliability.

Who this book is for

The target audience for this book should have a good understanding of GeoServer use and web hosting.

What this book covers

Chapter 1, *Developing a Spatial Analysis Platform with WPS*, will teach you how to create web-processing services to expose your data for live geospatial processing by your end user.

Chapter 2, *Speeding Up Your App with Tile Caching*, will explain tile caching and how to use it to create an app, and will help you to resolve any issues that may occur in the process.

Chapter 3, *Optimizing GeoServer*, will consist of techniques for getting the best performance out of GeoServer through the use of data formats and clustering, and will help you to ensure reliable performance for your end user.

Chapter 4, *Secure Authentication*, will teach you how to add secure authentication to your GeoServer resources with TLS and GeoServer's authentication management system.

Chapter 5, *Putting it into Production*, will consist of the final steps in putting your app into production, including hosting options and management tasks.

To get the most out of this book

The main technical requirement is a PC running Windows, Linux, or macOS. Your PC should be running an OS from the last 5 years. The emphasis is on Windows with coverage of Linux, but Mac users will find the GUI approach in Windows most applicable to their own environment.

Download the example code files

You can download the example code files for this book from your account at www.packtpub.com. If you purchased this book elsewhere, you can visit www.packtpub.com/support and register to have the files emailed directly to you.

You can download the code files by following these steps:

1. Log in or register at www.packtpub.com.
2. Select the **SUPPORT** tab.
3. Click on **Code Downloads & Errata**.
4. Enter the name of the book in the **Search** box and follow the onscreen instructions.

Once the file is downloaded, please make sure that you unzip or extract the folder using the latest version of:

- WinRAR/7-Zip for Windows
- Zipeg/iZip/UnRarX for Mac
- 7-Zip/PeaZip for Linux

The code bundle for the book is also hosted on GitHub at **https://github.com/PacktPublishing/Expert-GeoServer**. In case there's an update to the code, it will be updated on the existing GitHub repository.

We also have other code bundles from our rich catalog of books and videos available at https://github.com/PacktPublishing/. Check them out!

Download the color images

We also provide a PDF file that has color images of the screenshots/diagrams used in this book. You can download it here: https://www.packtpub.com/sites/default/files/downloads/ExpertGeoServer_ColorImages.pdf.

Conventions used

There are a number of text conventions used throughout this book.

`CodeInText`: Indicates code words in text, database table names, folder names, filenames, file extensions, pathnames, dummy URLs, user input, and Twitter handles. Here is an example: "The `httpd` file includes some conditional directives for loading modules."

A block of code is set as follows:

```
<wps:Data>
  <wps:LiteralData>Min</wps:LiteralData>
</wps:Data>
```

When we wish to draw your attention to a particular part of a code block, the relevant lines or items are set in bold:

```
projection: new ol.proj.Projection({
  code: 'EPSG:4326',
  units: 'degrees',
  axisOrientation: 'neu',
  global: true
})
```

Any command-line input or output is written as follows:

```
openssl req -x509 -nodes -days 365 -newkey rsa:2048 -keyout
C:\Apache24\conf\ssl\geoserver-demo.key -out
C:\Apache24\conf\ssl\geoserver-demo.crt
```

Bold: Indicates a new term, an important word, or words that you see onscreen. For example, words in menus or dialog boxes appear in the text like this. Here is an example: "Select the **Execute process** button, and you can see a successful output."

Warnings or important notes appear like this.

Tips and tricks appear like this.

Get in touch

Feedback from our readers is always welcome.

General feedback: Email `feedback@packtpub.com` and mention the book title in the subject of your message. If you have questions about any aspect of this book, please email us at `questions@packtpub.com`.

Errata: Although we have taken every care to ensure the accuracy of our content, mistakes do happen. If you have found a mistake in this book, we would be grateful if you would report this to us. Please visit `www.packtpub.com/submit-errata`, selecting your book, clicking on the Errata Submission Form link, and entering the details.

Piracy: If you come across any illegal copies of our works in any form on the Internet, we would be grateful if you would provide us with the location address or website name. Please contact us at `copyright@packtpub.com` with a link to the material.

If you are interested in becoming an author: If there is a topic that you have expertise in and you are interested in either writing or contributing to a book, please visit `authors.packtpub.com`.

Reviews

Please leave a review. Once you have read and used this book, why not leave a review on the site that you purchased it from? Potential readers can then see and use your unbiased opinion to make purchase decisions, we at Packt can understand what you think about our products, and our authors can see your feedback on their book. Thank you!

For more information about Packt, please visit `packtpub.com`.

Developing a Spatial Analysis Platform with WPS

1

In this chapter, you'll learn how to use the **Web Processing Service** (**WPS**) to run geospatial processing on your data and return the results to your web app. We'll start out by installing the WPS plugin on your GeoServer instance, and you'll learn how it works. Then you'll use the WPS request builder to create some simple WPS requests. You'll explore the process chaining technique with a more complex example, and finally, we'll integrate the complex WPS request into OpenLayers.

In this chapter, we will cover the following topics:

- Installing and learning the basics of WPS
- WPS request builder
- Process chaining
- OpenLayers integration

Installing and learning the basics of WPS

Let's get started by installing and learning the basics of WPS. We'll start by installing the plugin and noticing changes in the web administration interface within GeoServer. You'll learn about the WPS standard and basic structure, and finally, we'll look at a simple WPS request.

WPS is an interface standard, similar to WMS and WFS; however, the data output by WPS is dynamically created by a process or set of processes run by the WPS server; in our case, GeoServer. Like WFS, a request can be sent synchronously or asynchronously via HTTP POST or HTTP GET. The WPS standard defines the format of the expected input and output of the processes it interfaces with, so the data that you send in needs to match the expected data format of the service.

An example of this could be a buffer process that takes, as an input, a geographical coordinate expressed as JSON, and produces a polygon overlay that is perhaps produced as GML, and then that is displayed on a map in a web app, as shown in the following diagram:

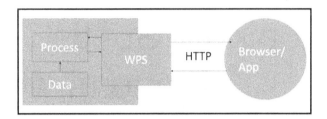

First we need to install the WPS plugin with GeoServer, as this capability is not available in GeoServer without the plugin. The way that you do that is the same as for other plugins and GeoServer. You go to the released download page (`http://geoserver.org/release/stable/`), and the **Extensions** section is at the bottom. You download the WPS services zip archive, and expand it, and then pull it into the `WEB-INF` plugins directory. WPS can be used with HTTP GET and HTTP POST.

The following is an HTTP GET WPS request:

```
http://localhost:8080/geoserver/ows?service=wps&version=1.0.0&request=GetCa
pabilities
```

After we've installed the WPS plugin, we can now send a WPS request to GeoServer, and, through this URL, we're using the get capabilities operation within WPS which, like WFS or WMS, just gives us an XML representation of all of the processes that the WPS server supports.

And you can see, here we have many processes, such as this envelope process, which you may recognize from other GIS packages you might have used, as shown in the following code snippet:

```
<wps:Process wps:processVersion="1.0.0">
  <ows:Identifier>geo:envelope</ows:Identifier>
  <ows:Title>Envelope</ows:Title>
  <ows:Abstract>
    Returns the smallest bounding box polygon that contains a geometry.
    For a point geometry, returns the same point.
  </ows:Abstract>
</wps:Process>
```

Here's a polygon extraction process, as shown in the following code snippet, and these are namespaces that correspond to the underlying code that's used to create the WPS service:

```
<wps:Process wps:processVersion="1.0.0">
  <ows:Identifier>ras:PolygonExtraction</ows:Identifier>
  <ows:Title>Polygon Extraction</ows:Title>
  <ows:Abstract>
    Extracts vector polygons from a raster, based on regions which are
    equal or in given ranges
  </ows:Abstract>
</wps:Process>
```

This is created before we install the plugin. It's more or less outside of our control, unless we want to do coding in Java, I guess. So, there are some different namespaces here, such as `ras:PolygonExtraction`, `vec:VectorZonalStatistics`, and so on. These are not something you'll be modifying, really, but it's good to know that there are different operations within different namespaces in WPS.

After we've installed the WPS plugin, we also have some additional options in our GeoServer instance. This process status is connected to WPS. We have a WPS **Settings** area, and a **Security** area, and in our **Demos** we now have the ability to create requests with a **WPS request builder** and **Demo requests**.

Just a word about WPS syntax: the syntax, similarly to when we were using WFS and WMS, is defined on a specification, and you can get the details of the specification if you go through the documentation. Open Geospatial (`http://docs.opengeospatial.org/is/14-065/14-065.html#15`) is the best place to look for that. You can see that this execution type request (WPS `GetCapabilities` request), as shown in the following diagram, will be the main operation that we'll be using to run different kinds of processes:

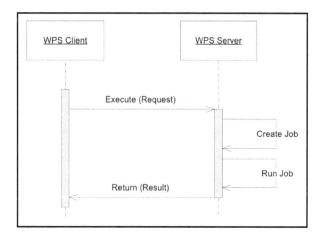

The processes are those geometric operations that we noticed, and they get a capabilities document response. And you'll get a better example of how an execute operation is defined in the WPS XML, but I will mention that you'll see the identifier or come up a lot, which can be used to identify a process or identify a reference, which will be usually some kind of external data. We'll also see some kind of XML about the response and how to format that.

So, let's take a look at a simple WPMS, or WPS request, through the **Demo requests** area of GeoServer. We've used this in the past with WFST, and we have these WPS options now, after installing WPS, as shown in the following screenshot:

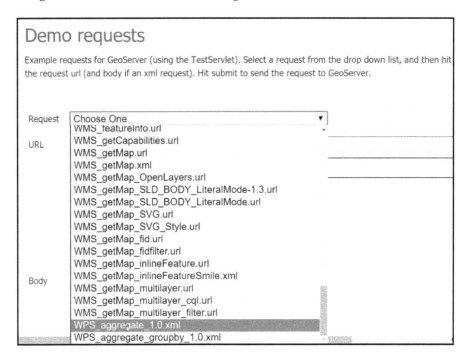

Let's try to check out a WPS request. The following code will be sent as a HTTP POST request via the following XML:

```xml
<?xml version="1.0" encoding="UTF-8"?><wps:Execute version="1.0.0"
service="WPS" xmlns:xsi="http://www.w3.org/2001/XMLSchema-instance"
xmlns="http://www.opengis.net/wps/1.0.0"
xmlns:wfs="http://www.opengis.net/wfs"
xmlns:wps="http://www.opengis.net/wps/1.0.0"
xmlns:ows="http://www.opengis.net/ows/1.1"
xmlns:gml="http://www.opengis.net/gml"
xmlns:ogc="http://www.opengis.net/ogc"
xmlns:wcs="http://www.opengis.net/wcs/1.1.1"
xmlns:xlink="http://www.w3.org/1999/xlink"
```

```
xsi:schemaLocation="http://www.opengis.net/wps/1.0.0
http://schemas.opengis.net/wps/1.0.0/wpsAll.xsd">
  <ows:Identifier>gs:Aggregate</ows:Identifier>
  <wps:DataInputs>
    <wps:Input>
      <ows:Identifier>features</ows:Identifier>
      <wps:Reference mimeType="text/xml"
       xlink:href="http://geoserver/wfs" method="POST">
        <wps:Body>
          <wfs:GetFeature service="WFS" version="1.0.0"
           outputFormat="GML2"
           xmlns:sf="http://www.openplans.org/spearfish">
            <wfs:Query typeName="topp:states"/>
          </wfs:GetFeature>
        </wps:Body>
      </wps:Reference>
    </wps:Input>
    <wps:Input>
      <ows:Identifier>aggregationAttribute</ows:Identifier>
      <wps:Data>
        <wps:LiteralData>PERSONS</wps:LiteralData>
      </wps:Data>
    </wps:Input>
    <wps:Input>
      <ows:Identifier>function</ows:Identifier>
      <wps:Data>
        <wps:LiteralData>Count</wps:LiteralData>
      </wps:Data>
    </wps:Input>
    <wps:Input>
      <ows:Identifier>function</ows:Identifier>
      <wps:Data>
        <wps:LiteralData>Average</wps:LiteralData>
      </wps:Data>
    </wps:Input>
    <wps:Input>
      <ows:Identifier>function</ows:Identifier>
      <wps:Data>
        <wps:LiteralData>Sum</wps:LiteralData>
      </wps:Data>
    </wps:Input>
    <wps:Input>
      <ows:Identifier>function</ows:Identifier>
      <wps:Data>
        <wps:LiteralData>Min</wps:LiteralData>
      </wps:Data>
    </wps:Input>
    <wps:Input>
```

```
        <ows:Identifier>function</ows:Identifier>
        <wps:Data>
          <wps:LiteralData>Max</wps:LiteralData>
        </wps:Data>
      </wps:Input>
      <wps:Input>
        <ows:Identifier>singlePass</ows:Identifier>
        <wps:Data>
          <wps:LiteralData>false</wps:LiteralData>
        </wps:Data>
      </wps:Input>
    </wps:DataInputs>
    <wps:ResponseForm>
      <wps:RawDataOutput mimeType="application/json">
        <ows:Identifier>result</ows:Identifier>
      </wps:RawDataOutput>
    </wps:ResponseForm>
  </wps:Execute>
```

You can see `Identifier` identifying the `Aggregate` operation or process, and `gs` is the namespace. This just means that this is a GeoServer-specific WPS process, and there will be some features that are given as input into this process. In this case, the features are coming from the WFS service on the GeoServer, and there are a couple of other parameters that are defined here in input and identifier areas, along with literal data. Here you can see the output format; the raw data output will be expressed as JSON. So now, if we **Submit** this request, in return we will get the aggregation results expressed in JSON, which is as follows:

```
{"GroupByAttributes":[],"AggregationResults":
[[2.46881454E8,49,453588,5038397.020408163,2.9760021E7]],"AggregationFunctions
":["Sum","Count","Min","Average","Max"],"AggregationAttribute":"PERSONS"}
```

This is aggregating some information about states and population. So now, you've seen a simple WPS request. In the next section, you'll see how to create a WPS request with a request builder.

WPS request builder

In the previous section, you installed and learned the basics of WPS. In this section, you'll learn about the GeoServer WPS request builder tool in the web administration interface, and how you can use it to generate requests based on WPS processes and parameters. Throughout this section, we'll create a single request in the builder. We'll look at processes and select process. Next, we'll consider and select inputs, layer, reference, text, and subprocess, and view the constructed XML request. Finally, we'll select a format for output and examine the results. As discussed in the previous section, the request builder is accessed from the demos page. The first step is to choose a process, just as shown in the following screenshot:

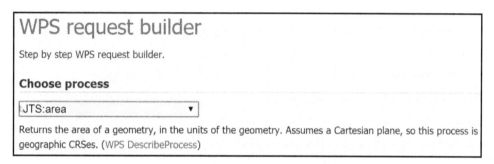

JTS refers to the **Java Topology Suite** namespace, I believe, and this area process is just a process that happens to be attached to that package; it's part of GeoServer. **JTS:area** returns the area of a geometry, in the units of the geometry, as shown in the preceding screenshot, so the input here is a geometry.

The geometry can be directly input as XML, GML, or any of the following formats:

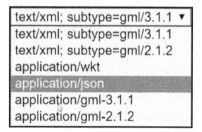

Another option would be to **REFERENCE** the geometry. You can get it from GeoServer, as follows:

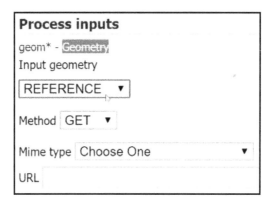

You just need to put a URL in there to get the correct geometry.

With all of these different options, it's best to break the problem down into smaller units, so you can test and make sure you're getting the output that you're expecting.

There's an option in here, **SUBPROCESS**, which I've found to not work. Theoretically, you could use some kind of process chaining in here; for example, you could use `gs:CollectGeometries`, which turns a collection into a geometry, and then the geometry would be useful for that area process. Unfortunately, I have not had any success with this, so I'm not going to get into using that subprocess here to create process chaining in the builder, but I will say that you can select a subprocess and, sometimes, you can get some kind of XML output.

One of my favorite things to do with the request builder is to just get XML output. For instance, we can see what the actual request would look like as created by the request builder, as shown in the following code, and then copy and paste that into an XML document, edit it, and make it actually work:

```
<?xml version="1.0" encoding="UTF-8"?><wps:Execute version="1.0.0"
service="WPS" xmlns:xsi="http://www.w3.org/2001/XMLSchema-instance"
xmlns="http://www.opengis.net/wps/1.0.0"
xmlns:wfs="http://www.opengis.net/wfs"
xmlns:wps="http://www.opengis.net/wps/1.0.0"
xmlns:ows="http://www.opengis.net/ows/1.1"
xmlns:gml="http://www.opengis.net/gml"
xmlns:ogc="http://www.opengis.net/ogc"
xmlns:wcs="http://www.opengis.net/wcs/1.1.1"
xmlns:xlink="http://www.w3.org/1999/xlink"
xsi:schemaLocation="http://www.opengis.net/wps/1.0.0
```

```
http://schemas.opengis.net/wps/1.0.0/wpsAll.xsd">
  <ows:Identifier>JTS:area</ows:Identifier>
  <wps:DataInputs>
    <wps:Input>
      <ows:Identifier>geom</ows:Identifier>
      <wps:Data>
        <wps:ComplexData mimeType="text/xml;
          subtype=gml/3.1.1"><![CDATA[Some text]]>
        </wps:ComplexData>
      </wps:Data>
    </wps:Input>
  </wps:DataInputs>
  <wps:ResponseForm>
    <wps:RawDataOutput>
      <ows:Identifier>result</ows:Identifier>
    </wps:RawDataOutput>
  </wps:ResponseForm>
</wps:Execute>
```

Let's try that `gs:CollectGeometries` process under the **Choose process** dropdown, because I know that we can use an existing GeoServer layer there, and we can view the output. Select `tiger:giant_polygon` from the **Process inputs** dropdown.

There are a couple of different options for output; let's get it in JSON by selecting `application/json` from the **Process outputs** dropdown.

Select the **Execute process** button, and you can see successful output, as shown in the following screenshot:

```
{"type":"MultiPolygon","coordinates":[[[[-180,-90],[-180,90],[180,90],
[180,-90],[-180,-90]]]]}
```

So, this is useful as a kind of testing tool to build up parts of your process if you're creating a complex process with process chaining, which we'll be learning about in the next section. If you want to preview the GeoJSON, one nice site to do it at is `http://geojsonlint.com/`, and you just paste in your GeoJSON. If it's valid, it'll show up on the map, as shown in the following screenshot; if it's not, it'll give you a pretty good error message about why it's not working:

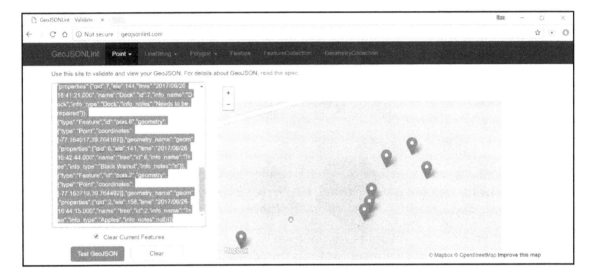

Also, QGIS is always helpful when you're going through a process and are able to preview data output.

In this section, you learned about building a WPS request with request builder, and about some of the parts of a WPS request, especially input formats. In the next section, you'll learn more about building a complex request with process chaining, and we can bring some of the XML we created with our request builder into the more complex process-chained WPS request.

Process chaining

In the previous section, you learned about creating WPS requests with WPS request builder. In this section, you'll learn how to make more complex WPS requests with process chaining. First, we'll take a look at a chain process request in plain XML. Next, we'll create an even more complex request. Finally, we'll add this request to a minimal OpenLayers web app to test the results of POST.

Here is our first process-chained XML, and you can see that the first WPS process is started here with this `wps:Execute` operation. The identifier is `gs:CollectGeometries`, so it's the `gs:CollectGeometries` process that we're running here, as shown in the following code:

```xml
<?xml version="1.0" encoding="UTF-8"?>
<wps:Execute version="1.0.0" service="WPS"
xmlns:xsi="http://www.w3.org/2001/XMLSchema-instance"
xmlns="http://www.opengis.net/wps/1.0.0"
xmlns:wfs="http://www.opengis.net/wfs"
xmlns:wps="http://www.opengis.net/wps/1.0.0"
xmlns:ows="http://www.opengis.net/ows/1.1"
xmlns:gml="http://www.opengis.net/gml"
xmlns:ogc="http://www.opengis.net/ogc"
xmlns:wcs="http://www.opengis.net/wcs/1.1.1"
xmlns:xlink="http://www.w3.org/1999/xlink"
xsi:schemaLocation="http://www.opengis.net/wps/1.0.0
http://schemas.opengis.net/wps/1.0.0/wpsAll.xsd">
  <ows:Identifier>gs:CollectGeometries</ows:Identifier>
  <wps:DataInputs>
    ...
    // Full code in the code bundle
  </wps:ResponseForm>
</wps:Execute>
```

But the interesting thing is that when you look at `wps:Input` of `CollectGeometries`, you can see it's getting the `features` parameter from `wps:Reference`. So, this is not `wps:Execute`, yet it still is `wps:Reference`, but it's referencing a `text/xml` type that includes an `wps:Execute` operation in it, which is a little bit convoluted, but that's just how it's set up. And this is usually how you're going to refer to process-chained processes, with XML inside of other processes.

So, we have our second `wps:Execute` operation here, inside a `wps:Body` section, and this is using the `BufferFeatureCollection` process, which also has a `features` input parameter:

```xml
<wps:Body>
  <wps:Execute version="1.0.0" service="WPS">
   <ows:Identifier>gs:BufferFeatureCollection</ows:Identifier>
    <wps:DataInputs>
      <wps:Input>
        <ows:Identifier>features</ows:Identifier>
        <wps:Reference mimeType="text/xml"
xlink:href="http://geoserver/wfs" method="POST">
          <wps:Body>
            <wfs:GetFeature service="WFS" version="1.0.0"
```

```
outputFormat="GML2" xmlns:learning-
geoserver="http://packtpub.com/learning-geoserver">
              <wfs:Query srsName="EPSG:2272" typeName="learning-
geoserver:pois"/>
            </wfs:GetFeature>
          </wps:Body>
        </wps:Reference>
      </wps:Input>
```

Here, for `features`, we're also using `wps:Reference`, but rather than having an additional WPS request or WPS section inside of `wps:Reference`, we're using `wfs`. This section in here is actually just a valid `wfs` defined by the WFS standard:

```
<wfs:GetFeature service="WFS" version="1.0.0" outputFormat="GML2"
xmlns:learning-geoserver="http://packtpub.com/learning-geoserver">
  <wfs:Query srsName="EPSG:2272" typeName="learning-geoserver:pois"/>
</wfs:GetFeature>
```

So anything that you would put, for example, in a `wfs:Query` section on WFS would be valid, and in this section of this XML. We're just using a regular `wfs:Query`, getting all of the points of interest in the `learning-geoserver:pois` workspace. Some things to be aware of here are, again, data types, so we're going to be outputting the following little portion as text XML-based `wfs-collection`:

```
<wps:RawDataOutput mimeType="text/xml; subtype=wfs-collection/1.0">
  <ows:Identifier>result</ows:Identifier>
</wps:RawDataOutput>
```

That is just the way this buffer features collection process returns data, and, of course, `gs:CollectGeometries` will turn that collection into geometries. So, be aware of data formats. You'll be using a lot of different formats through a process-chained operation. A lot of the time, you'll be changing between different formats, so of course, the spatial reference will be important. In this case, we'd have trouble doing some of these operations if we're not using a linear spatial reference, and we end up having to change or reproject the spatial reference into a different system later on, which you'll see in the next section, with OpenLayers.

The namespaces are also very important. Because this is just a feature of XML, you have to explain what your tag is referring to so that all the different data types appear in their own way. If you are using XML namespaces correctly, it separates out portions of the XML in the correct way so that the services you're feeding it to understand how to read the data.

You can see many namespaces are defined up here; not all of them are necessary, but you can see that `wfs` is there, defining this little section and the syntax for this section, which comes from a WFS standard:

```
<wps:Reference mimeType="text/xml" xlink:href="http://geoserver/wfs"
method="POST">
  <wps:Body>
    <wfs:GetFeature service="WFS" version="1.0.0" outputFormat="GML2"
xmlns:learning-geoserver="http://packtpub.com/learning-geoserver">
      <wfs:Query srsName="EPSG:2272" typeName="learning-geoserver:pois"/>
    </wfs:GetFeature>
  </wps:Body>
</wps:Reference>
```

Here is the result of this as GML:

```
<wps:RawDataOutput mimeType="text/xml; subtype=gml/3.1.1">
  <ows:Identifier>result</ows:Identifier>
</wps:RawDataOutput>
```

So, that's our first XML example of a WPS request. Let's look at our second example, as follows:

```
<ows:Identifier>JTS:difference</ows:Identifier>
  <wps:DataInputs>
    <wps:Input>
      <ows:Identifier>a</ows:Identifier>
      <wps:Reference mimeType="text/xml; subtype=gml/3.1.1"
xlink:href="http://geoserver/wps" method="POST">
        <wps:Body>
          <wps:Execute version="1.0.0" service="WPS">
            <ows:Identifier>gs:CollectGeometries</ows:Identifier>
            <wps:DataInputs>
              <wps:Input>
                <ows:Identifier>features</ows:Identifier>
                <wps:Reference mimeType="text/xml"
xlink:href="http://geoserver/wps" method="POST">
```

This is our second WPS request, and it will be even more complex than the last one, with process chaining. In the last process chain, we look at this `gs:CollectGeometries` section and everything below it. In this process-chained operation, we're using this `difference` process, which is based on a `JTS` namespace, and this takes two `wps:Input` parameters:

```
<ows:Identifier>JTS:difference</ows:Identifier>
```

Each of these inputs gives geometry in GML, and this process will take the difference of these geometries and give us that as a separate geometry feature collection or single geometry in GeoJSON format:

```
<wps:RawDataOutput mimeType="application/json">
  <ows:Identifier>result</ows:Identifier>
</wps:RawDataOutput>
```

Now I'm moving over into OpenLayers. We can run a very simple WPS request as follows:

```
<!DOCTYPE html>
<html lang="en">
<head>
  <title>WPS-Request Example</title><!-- Required meta tags -->
  <meta charset="utf-8">
  <meta name="viewport" content=
 "width=device-width, initial-scale=1, shrink-to-fit=no">
  <!-- OL stylesheet -->
  <link rel="stylesheet" href="https://openlayers.org/en/v4.4.2/css/ol.css"
type="text/css">
</head>
<body>
  <h1>WPS-Request Example</h1>

<!-- OpenLayers JS dependency, debug build -->
<script type="text/javascript"
src="https://openlayers.org/en/v4.4.2/build/ol-debug.js"></script>
<!-- Begin wps-request Javascript -->
<script type="text/javascript">
```

The preceding code is based on a stripped-down OpenLayers app, which sends an HTTP POST request with XML. We can start with this, doing incremental testing to build up a chained WPS request in the OpenLayers code. So, it's pretty stripped down; there's not a whole lot going on there. We just have our `postData` with all of the WPS text XML body, which is as follows:

```
postData = '<?xml version="1.0" encoding="UTF-8"?><wps:Execute
version="1.0.0" service="WPS"
xmlns:xsi="http://www.w3.org/2001/XMLSchema-instance"
xmlns="http://www.opengis.net/wps/1.0.0"
xmlns:wfs="http://www.opengis.net/wfs"
xmlns:wps="http://www.opengis.net/wps/1.0.0"
xmlns:ows="http://www.opengis.net/ows/1.1"
xmlns:gml="http://www.opengis.net/gml"
xmlns:ogc="http://www.opengis.net/ogc"
xmlns:wcs="http://www.opengis.net/wcs/1.1.1"
xmlns:xlink="http://www.w3.org/1999/xlink"
```

```
xsi:schemaLocation="http://www.opengis.net/wps/1.0.0
http://schemas.opengis.net/wps/1.0.0/wpsAll.xsd">
<ows:Identifier>gs:Aggregate</ows:Identifier><wps:DataInputs><wps:Input><ow
s:Identifier>features</ows:Identifier><wps:Reference mimeType="text/xml"
xlink:href="http://geoserver/wfs" method="POST"><wps:Body><wfs:GetFeature
service="WFS" version="1.0.0" outputFormat="GML2"
xmlns:sf="http://www.openplans.org/spearfish"><wfs:Query
typeName="topp:states"/></wfs:GetFeature></wps:Body></wps:Reference></wps:I
nput><wps:Input><ows:Identifier>aggregationAttribute</ows:Identifier><wps:D
ata><wps:LiteralData>PERSONS</wps:LiteralData></wps:Data></wps:Input><wps:I
nput><ows:Identifier>function</ows:Identifier><wps:Data><wps:LiteralData>Co
unt</wps:LiteralData></wps:Data></wps:Input><wps:Input><ows:Identifier>func
tion</ows:Identifier><wps:Data><wps:LiteralData>Average</wps:LiteralData></
wps:Data></wps:Input><wps:Input><ows:Identifier>function</ows:Identifier><w
ps:Data><wps:LiteralData>Sum</wps:LiteralData></wps:Data></wps:Input><wps:I
nput><ows:Identifier>function</ows:Identifier><wps:Data><wps:LiteralData>Mi
n</wps:LiteralData></wps:Data></wps:Input><wps:Input><ows:Identifier>functi
on</ows:Identifier><wps:Data><wps:LiteralData>Max</wps:LiteralData></wps:Da
ta></wps:Input><wps:Input><ows:Identifier>singlePass</ows:Identifier><wps:D
ata><wps:LiteralData>false</wps:LiteralData></wps:Data></wps:Input></wps:Da
taInputs><wps:ResponseForm><wps:RawDataOutput
mimeType="application/json"><ows:Identifier>result</ows:Identifier></wps:Ra
wDataOutput></wps:ResponseForm></wps:Execute>';
```

Then, `postData` is sent via this XML HTTP request object posted to GeoServer. Then we'll get a response in as an alert, which can be achieved with the following code:

```
url = 'http://localhost:8080/geoserver/wps';
var req = new XMLHttpRequest();
req.open("POST", url, true);
req.setRequestHeader('Content-type', 'text/xml');
req.onreadystatechange = function() {
  if (req.readyState != 4) return;
  if (req.status != 200 && req.status != 304) {
    alert('HTTP error ' + req.status);
    return;
  }
  alert(req.responseText);
  if (req.readyState == 4) return;
};
req.send(postData);
```

So, we can test that out, and you can see the following result of that operation, which is in JSON; it's the data corresponding to the request:

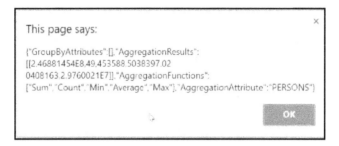

Taking that one step further, we just add the XML from the more complex chained operation that we showed in the second example XML previously. Now, if we run this process-chained POST request, running this more complex process-chained POST request will sometimes require some time before it returns a result. Here you can see the result as GeoJSON:

Again, the spatial reference system is linear, so we're seeing much longer coordinates than we would normally see, but we're getting a successful result from our complex process-chained operation.

You've seen now how to create a complex process-chained operation. In the next section, we'll look at full integration of that WPS request into OpenLayers, as well as the result.

OpenLayers integration

In the previous section, you learned how to create and test complex WPS requests. In this section, you'll round off your understanding of WPS by fully integrating requests and responses with OpenLayers. We'll be using the example of determining where to plant apple trees on a farm using some of the data that we've worked with in previous sections. We will first cover the GeoJSON OpenLayers class, and then we'll look at how the WPS response is read into OpenLayers with that. Then we'll cover reprojection of the data with Proj4JS in particular, and finally, we'll view the result. You can refer to the following documentation for more information:

```
http://openlayers.org/en/latest/apidoc/module-ol_format_GeoJSON-GeoJSON.
html#writeFeatureObject
```

We're going to use the GeoJSON class to read data into our OpenLayers app, and it's useful here to take a look at some of the methods that are available, such as `readFeature` and `readFeatures`. That distinction is important, since `readFeatures` takes an array of OpenLayers features that consists of multiple features:

```
readFeatures(source, opt_options) -> {Array.<module:ol/Feature~Feature>}
```

On the other hand, `readFeature` only takes a single feature:

```
readFeature(source, opt_options) -> {module:ol/Feature~Feature}
```

You might want to check out some of these other methods as well. Let's take a look at the code. We have our app example that we were working with in the previous section, and we've now added a dependency for Proj4JS by adding the following line:

```
<script
src="https://cdnjs.cloudflare.com/ajax/libs/proj4js/2.4.4/proj4.js"></scrip
t>
```

PROJ, by the way, is the standards-based, open source-based way of doing projections by reprojecting data, and it's used by many different software packages. This Proj4JS, of course, is the JavaScript version of PROJ, and since we're working with EPSG 2272 data in our WPS operation so that we can do some linear-based operations, we need to reproject that data as `EPSG: 4326`, compatible with OpenLayers.

The first thing we need to do is tell PROJ how to handle this data format, and you can find a `proj4` string at `http://spatialreference.org/`, which is a great source, but a PROJ string is a standardized way where PROJ can know some details of the coordinate system or projection system, and that way, it can translate between different projections and coordinate systems:

```
proj4.defs('EPSG:2272', '+proj=lcc +lat_1=40.96666666666667
+lat_2=39.93333333333333 +lat_0=39.33333333333334 +lon_0=-77.75 +x_0=600000
+y_0=0 +ellps=GRS80 +datum=NAD83 +to_meter=0.3048006096012192 +no_defs');
var proj2272 = ol.proj.get('EPSG:2272');
```

So, we've taken care of some PROJ details here, creating this new variable, which is our PROJ-defined projection system. Our view is using `EPSG: 4326`:

```
var view = new ol.View({
  center: [-77.163785, 39.7641],
  zoom: 19,
  projection: new ol.proj.Projection({
    code: 'EPSG:4326',
    units: 'degrees',
    axisOrientation: 'neu',
    global: true
  })
});
```

If we don't use `4326` or `3857`, we won't be able to use the OpenStreetMap source-based map. We're also pulling in a WMS from our GeoServer instance, which is the points of interest that we worked with previously, and that'll give us a bit of a reference for the result of this WPS operation:

```
var wms = new ol.layer.Image({
  //extent: [-13884991, 2870341, -7455066, 6338219],
  source: new ol.source.ImageWMS({
    url: host + '/wms',
    params: {'LAYERS': 'learning-geoserver:pois'},
    ratio: 1,
    serverType: 'geoserver'
  })
});
```

```
var map = new ol.Map({
  target: "map",
  layers: [osm, wms],
  view: view
});
```

The user will click on a button, which will go to the submit function, and so it will submit the WPS request, which is what we covered in the previous section. It's in XML format, and it will be posted to the server. Upon return, with our response, it will be reading that into a GeoJSON object. Importantly, here, you can see that dataProjection is in this proj2272 system, while we want it to be projected into EPSG:4326. So, there'll be a coordinate system translation going on there:

```
url = host + '/wps';
var req = new XMLHttpRequest();
req.open("POST", url, true);
req.setRequestHeader('Content-type', 'text/xml');
req.onreadystatechange = function() {
  if (req.readyState != 4) return;
  if (req.status != 200 && req.status != 304) {
    alert('HTTP error ' + req.status);
    return;
  }
  var format = new ol.format.GeoJSON();
  response = req.response;
  var feature = (format.readFeatures(response, {
    dataProjection: proj2272,
    featureProjection: 'EPSG:4326'
  }));
  var vectorSource = new ol.source.Vector({
    features: feature
  });
  var vectorLayer = new ol.layer.Vector({
    source: vectorSource
  });
  // console.log((new
ol.format.GeoJSON()).writeFeatures(vectorLayer.getSource().getFeatures()));
  map.addLayer(vectorLayer);
  if (req.readyState == 4) return;
};
req.send(postData);
}
```

We're just doing normal OpenLayers stuff in the preceding code, creating a `vectorSource` and a `vectorLayer` objects with our source set there. We've commented out this `console.log` line, but it's useful to do things like this for debugging. This will write the features of that `GeoJSON` object as GeoJSON, and we can use GeoJSON lint or QGIS to test and look at the result. The most important thing here is to notice whether the coordinates have been successfully reproductive, and then finally, with the `map` object, we'll just use the `addLayer` method to add `vectorLayer` to our map. So, let's take a look at this in action:

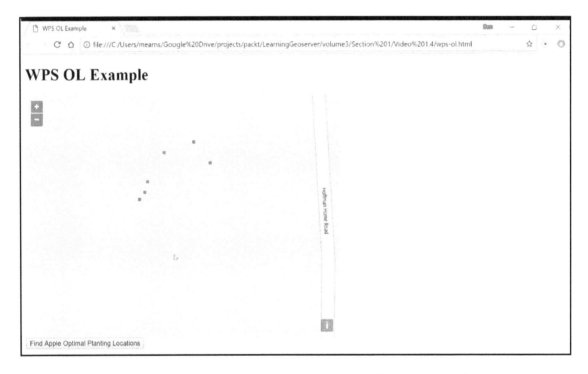

When you first bring up the app, you'll just see the points of interest and the base map. Now, to POST the WPS request, click the **Find Apple Optimal Planning Locations** button:

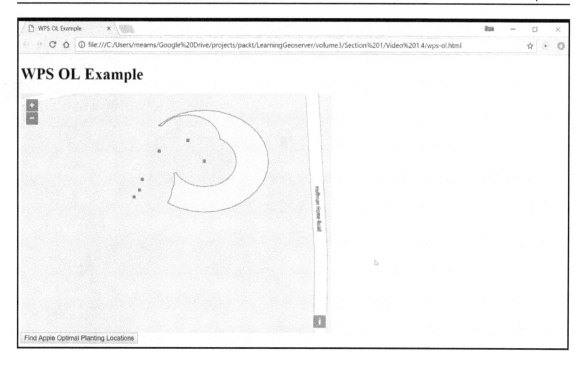

We want to find where the best place to plant apple trees is. Some of these trees are black walnut trees, which are not beneficial to apple trees and, in fact, may harm them, while some of these other trees are other apple trees, which we want to have close to the new apple trees for pollination purposes. And with WPS, we get this response, based on the location of the apple trees and the black walnut trees, of an area in which we can plant new apple trees.

Summary

In this chapter, we learned to use WPS to run geospatial processing on your data and return the results to our web app. We installed the WPS plugin on our GeoServer instance, and learned how it works. We used the WPS request builder to create some simple WPS requests. We also explored the process-chaining technique with a more complex example, and integrated the complex WPS request into OpenLayers.

In the next chapter, you'll learn about the technique of tile caching, and how you can harness tile caching to improve the performance of your app.

Speed Up Your App with Tile Caching
2

In this chapter, we will use tiles, pre-rendered portions of a map, to allow your app to load more rapidly. While the subject of tile caching can get pretty complex, I'll be covering the material you need to know to get your tile cache up and running and fix any issues that may occur.

First, we'll cover the tile caching basics. Then, you'll learn how to configure tile caching and GeoServer. Next, you'll create a tile-backed web map app, you'll learn to use the tile cache, and finally we'll explore methods for solving problems with tile caching.

In this chapter, we will cover the following topics:

- Tile caching basics
- Configure tile caching in GeoServer
- Creating a tile-backed OL app
- Using the tile cache
- Resolving problems with tile caching

Tile caching basics

In this section, you'll learn key concepts for tile caching, including tile caching schemes. We'll look at some software options for tile caching, and finally you'll learn about the process associated with the GeoServer tile caching stack that we'll be using in this section.

Tile caching is very useful when you are working with background maps, or for static content that does not change very often or involve much interaction. An example of tile caching is OpenStreetMap (`https://www.openstreetmap.org/#map=10/21.2516/86.7343`), and you can see tile caching at work as we zoom in and out of the map; portions of the map are pre-rendered as tiles, little images, and then they're stitched together by the frontend OpenLayers client, as shown in the following screenshot:

This allows very complex cartography to be delivered very quickly to our browser, avoiding the need to render that information in the browser or on the server in real time. Here, you can see two major title caching schemes in GeoServer (WMS-C service and TMS service), which are shown in the following screenshot:

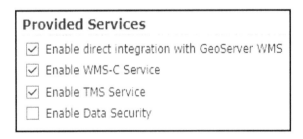

Tile caching schemes use different ways to refer to tiles and allow web clients to communicate with servers such as GeoServer to retrieve tiles in the appropriate way. While WMS-C is used for geographic coordinate bounding boxes to identify tiles, the new schemes, TMS and WMS-T, used coordinates based on the tile scheme itself, each tile numbered by its place on a grid, which is defined by variable grader origins. We can disable all of these other tile caching schemes as we are going to be using WMS-C, which is enabled separately, in its own section under **Services**.

The following example illustrates tile-based coordinates (`https://openlayers.org/en/latest/examples/canvas-tiles.html`):

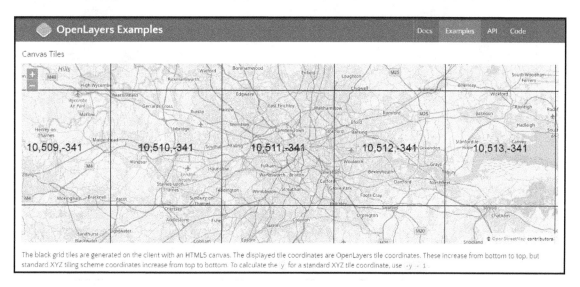

So, instead of seeing the typical latitude/longitude coordinates that you would be used to, you can see coordinates based on their place within the tile grid.

It's important to think about tile caches in terms of fixed parameters. Although there is some wiggle room for allowing the client to select variable parameters, you will get the best results by sticking to minimal parameters that fit your use case.

For example, in the case of image formats, select JPEG if compression is desired – this is a great option for imagery – and PNG if using vector data. Select PNG 8 if 256 colors are adequate, and simply PNG if a wider color palette is necessary; PNG 8 uses eight bits.

There are a number of software options for tile caching with your map app; these include GDAL or GDAL, the popular command-line geospatial utility Mapnik, which is the backend for the map box project and others, and GeoWebCache. GeoWebCache is available as a standalone instance, but is also embedded in GeoServer, so we will focus on GeoWebCache. Unlike file-based tile caches running on HTTP servers, GeoServer via GeoWebCache provides web service endpoints for map tiles for static and dynamic tile caching and additional capabilities.

The following diagram explains how tile caching works with GeoServer:

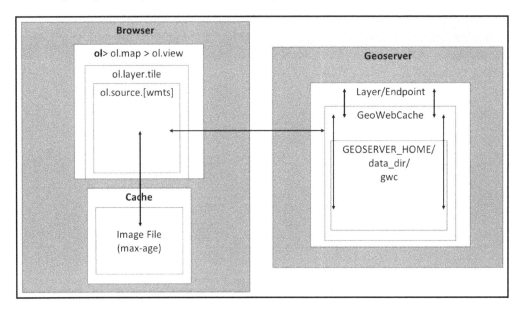

On the **Browser** side, we have a map client such as OpenLayers, and in the OpenLayers source there's a WMST class – our WMTS class that we can use for WMTS-type tile cache schemes. There are other classes available for other schemes. The OpenLayers app first checks on the browser cache to see if a tile is already available for the tile that's needed in the map view. Note that this max age of the **Image File** can be set in GeoServer to limit the caching of tiles. This is especially useful for debugging purposes.

On the **Geoserver** side, when a tile is not found in the cache, contact will be made with the GeoServer endpoint for that layer, which refers to **GeoWebCache**. It can directly refer to the **GeoWebCache** endpoint or to the WMS endpoint with a tiled parameter. We'll be using the direct **GeoWebCache** endpoint. When contact is made with the **GeoWebCache** endpoint, **GeoWebCache** will check on the filesystem to see if that image tile has already been created, and if so, that tile will be served back to the browser as a response to the HTTP request. If that tile is not already available, **GeoWebCache** will dynamically render that tile.

In this section, you learned the basics of tile caching. In the next section, you'll learn how to configure tile caching in GeoServer.

Configuring tile caching in GeoServer

In the last section, you learned the fundamentals of tile caching. In this section, you'll learn the ins and outs of setting up tile caching in GeoServer. First, you'll learn all the global and layer options for tile caching in the GeoServer web administration interface. Then, we'll look at these settings in the XML files under the GeoServer directory. Finally, you'll learn about gridsets and why these are important in building your tile-cache-backed app.

We disabled all of these other services in the previous section, as shown in the following screenshot, since we'll be using WMST, which is enabled in the **Services** portion of the sidebar:

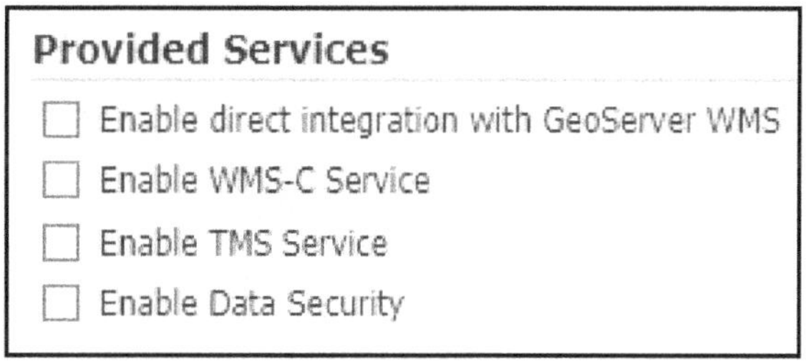

Provided Services

☐ Enable direct integration with GeoServer WMS

☐ Enable WMS-C Service

☐ Enable TMS Service

☐ Enable Data Security

You can set a number of global options that will be available on all new layers or layer groups that will be tiled. Let's look at a couple of these options:

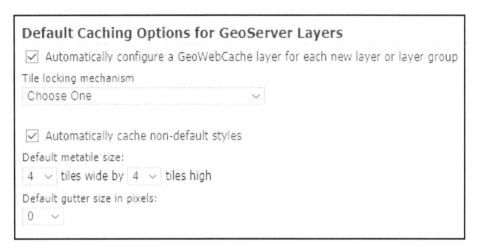

The metatile size and gutter size refer to ways of eliminating repeating labels or other tile artifacts that may appear during tile rendering. We'll talk more about this in the final section of this chapter.

There are a number of tile formats available; we covered this a little bit in the previous section, and we'll also cover this again in the last section of this chapter.

Another area to pay attention to here is the gridsets that are available, shown in the following screenshot, and we'll talk about those later in this section:

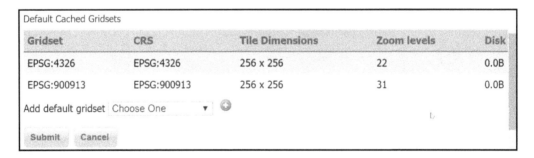

Formats are important for minimizing the disk space that is taken up, as well as improving the performance of GeoServer rendering and the transmission of tiles to the client app. Let's look at local options.

Options for particular layers or layer groups will also be defined on the layers themselves, or under the tile caching tab within a layer. As you can see in the following screenshot, caching is enabled for this layer:

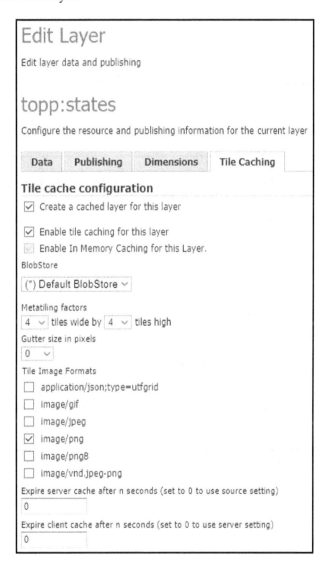

We're using the same default on the metatile. We have both of these formats, JPEG and PNG, enabled by default. A better option here would be to have just a single format selected, and we'll just select PNG, since this is a vector file format as shown in the preceding screenshot, as our vector data source.

An interesting area here is the **FORMAT_OPTIONS Parameter Filters** shown in the following screenshot, and these are additional URL parameters that will be allowed for this endpoint for the tile:

So, in this case, we needed the `dpi` parameter in our endpoint, and I'll show you how that's set up in XML. And the styles that are available for this layer are shown in the following screenshot. It's best to select only a single style and then disable alternative styles:

If alternative styles are enabled, tiles will be cached for those styles as well. If you don't need the additional gridsets, then I would recommend not enabling them for a particular tile endpoint:

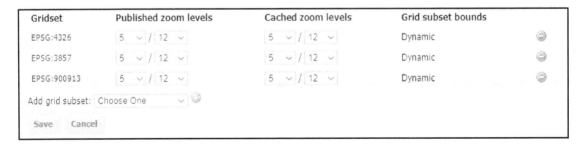

Now, let's take a look at the XML configuration that corresponds to the global and local web administration interface settings. So, we have the global XML for GeoWebCache, which can be found under the `data_dir\gwc` directory, and you can see that many of the same things that we just saw in the web administration interface are enabled here under the XML file. The reason that this is important is that the web administration interface sometimes fails to save settings.

I've had some problems with this in the past. I would recommend that, if you are doing anything complicated, you do it in the XML. You could try to start it on the web administration interface, and then, if you run into problems, you can complete it in the XML. You also might want to take backups of this, which you'll have if you are already backing up your data directory.

Here is an example of a local XML configuration for GeoWebCache. I want to show you, in particular, the `parameterFilters` section in the following code:

```
<parameterFilters>
  <stringParameterFilter>
    <key>FORMAT_OPTIONS</key>
    <defaultValue>dpi:72</defaultValue>
    <normalize>
      <locale></locale>
    </normalize>
    <values>
      <string>dpi:135</string>
      <string>dpi:90</string>
    </values>
  </stringParameterFilter>
  <styleParameterFilter>
    <key>STYLES</key>
    <defaultValue></defaultValue>
    <allowedStyles class="sorted-set"/>
    <availableStyles class="sorted-set">
      <string>polygon</string>
      <string>pophatch</string>
    </availableStyles>
    <defaultStyle>population</defaultStyle>
  </styleParameterFilter>
</parameterFilters>
```

And you can see how this format is laid out. We have a default option and a few strings that are available. If a value is outside of a certain range, it will not be accepted, but as we don't have any range set up here, we should be getting any value that is a string value; this value should be accepted for this format option dpi. This was necessary in one case where I had a tile that was an irregular size; however, you probably won't have to deal with that.

Gridsets are used for translating grid coordinates to geographic coordinates and vice versa. It's sometimes necessary to create a gridset for an unknown coordinate system by inputting information about the coordinate system. In particular, the gridset bounds and the tile width, **256**, will be the standard tile in pixels that we'll be working with:

Gridset bounds			
Min X	Min Y	Max X	Max Y
-20,037,508.34	-20,037,508.34	20,037,508.34	20,037,508.34

Compute from maximum extent of CRS

Tile width in pixels *

256

Tile height in pixels *

256

You can calculate the tile matrix set once you have the gridset bound set up and the tile width. I just added the zoom level 21 by clicking on the **Add zoom level** button, and that just automatically calculated the pixel size for that zoom level, as well as the scale denominator, and the name for this particular tile matrix would logically be EPSG:3857:21, but we'll take this out since we won't be using this:

14	9.554628534317017	1: 34,123.67333684649	EPSG:3857:14	16,384 x 16,384	⊖
15	4.777314267158508	1: 17,061.836668423246	EPSG:3857:15	32,768 x 32,768	⊖
16	2.388657133579254	1: 8,530.918334211623	EPSG:3857:16	65,536 x 65,536	⊖
17	1.194328566789627	1: 4,265.4591671058115	EPSG:3857:17	131,072 x 131,072	⊖
18	0.5971642833948135	1: 2,132.7295835529058	EPSG:3857:18	262,144 x 262,144	⊖
19	0.2985821416974068	1: 1,066.364791776453	EPSG:3857:19	524,288 x 524,288	⊖
20	0.1492910708487034	1: 533.1823958882266	EPSG:3857:20	1,048,576 x 1,048,576	⊖

Add zoom level

Save   Cancel

`EPSG:3857` is that web Mercator, of course, and it's identical to `EPSG:9100913`, and since that gridset is already available in GeoServer, this would be a good place to start if you want to just try to make a new gridset. You can pretty much copy and paste the values from there, or you can just add the zoom levels and it will populate the correct pixel size and scale. Another way to do this is to add it in the GeoWebCache XML that we were just looking at, the global XML, and that's an easier way to do it, really, because you can just copy and paste the plain text there in the XML.

In this section, you learned how to configure tile caching for GeoServer. In the next section, you'll learn how to create a tile-backed OpenLayers app.

Creating a tile-backed OL app

In the previous section, you learned how to configure GeoServer for tile caching. In this section, you will learn how to create a tile-backed OpenLayers app using the WMTS tile scheme. First, we'll look at a WMTS example in more detail. We'll examine the `CapabilitiesResponse` from our GeoServer endpoint, and finally we'll walk through the frontend code. WMTS is the current best practice for tile caching in GeoServer. You can think of it as the best of WMS and TMS, providing WMS-like syntax and capabilities with the tile grid coordinate performance of TMS. As with other OGC standards, the list of parameters for WMTS is available in the implementation standard document. The tile matrix set is based on pixel size, zoom levels, scale denominators, and bounding boxes. Essentially, more tiles will need to be generated at larger scales. You've really already seen this with the gridset example in the previous section.

You can see a tiled layer preview in the following screenshot; use WMTS for any layer by going to the **Tile Layers** section and selecting a gridset and format in the **Preview** dropdown:

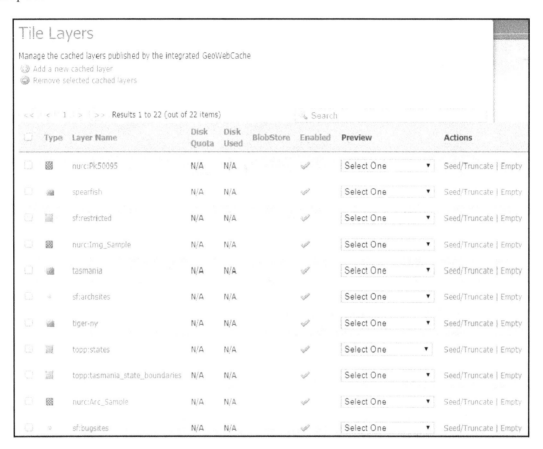

 So, we select the **tasmania** layer as an example and select **EPSG:3857/png** from the **Preview** drop-down menu. Once we get the PNG format image in the web browser, we look at the **Developer tools** to see a specific request and response. After zooming in one level, let's take a look at the developer tools of the browser. We're getting a Status 400 Bad Request, as shown in the following screenshot:

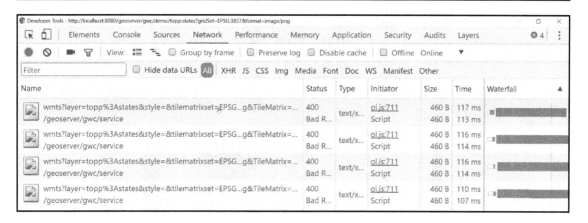

To get more information about this error, we could right-click and open a new tab for this particular resource. We will be getting a response exception in the web browser, `TileOutOfRange`, column 3 is out of range, `min: 4 max: 10`; this refers to the zoom level.

Let's zoom in a little closer. Now, we're starting to see some successful responses, as shown in the following screenshot:

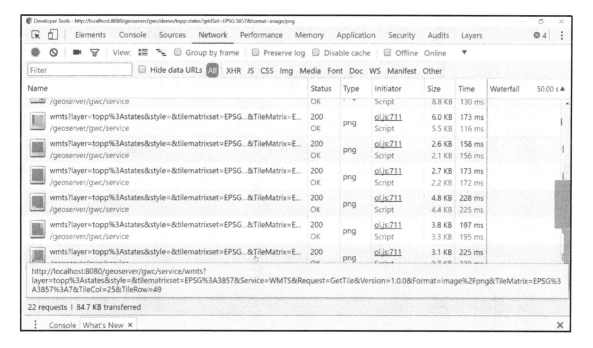

If we look at the URL for one of these responses, you'll notice that this URL has all the mandatory WMS parameters. For example, `TileRow`, `TileColumn`, and `TileMatrix` are specified, the request is of type `GetTile`, and the format is PNG.

We can take a look at a particular response and you'll see it's an actual image tile of type PNG:

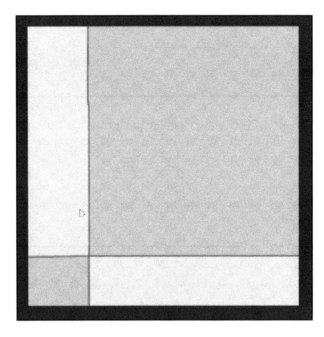

All of these PNGs are stitched together, of course, to make the map that looks like a smooth singular map. And there is also some transparency; the browser is showing us a kind of grid behind this to show that this is a transparent tile, so we could be layering tiles one on top of the other to allow different information to show through on the map.

Now, let's take a look at the code from the `wmts.html` file:

```
<!DOCTYPE html>
  <html>
    <head>
      <title>WMTS</title>
      <link rel="stylesheet"
href="https://openlayers.org/en/v4.6.4/css/ol.css" type="text/css">
      <!-- The line below is only needed for old environments like Internet
Explorer and Android 4.x -->
      <script
src="https://cdn.polyfill.io/v2/polyfill.min.js?features=requestAnimationFr
```

```
ame,Element.prototype.classList,URL"></script>
        <script
src="https://openlayers.org/en/v4.6.4/build/ol-debug.js"></script>
    </head>
    <body>
        <div id="map" class="map"></div>
        <script>
          var projection = ol.proj.get('EPSG:3857');
          var projectionExtent = projection.getExtent();
          var size = ol.extent.getWidth(projectionExtent) / 256;
          var resolutions = new Array(14);
          var matrixIds = new Array(14);
          for (var z = 0; z < 14; ++z) {
            // generate resolutions and matrixIds arrays for this WMTS
            resolutions[z] = size / Math.pow(2, z);
            matrixIds[z] = 'EPSG:3857:'+z;
          }

          var map = new ol.Map({
            layers: [
              new ol.layer.Tile({
                source: new ol.source.OSM(),
                opacity: 0.7
            }),
          new ol.layer.Tile({
              opacity: 0.7,
              source: new ol.source.WMTS({
                url: 'http://localhost:8080/geoserver/gwc/service/wmts',
                layer: 'topp:states',
                matrixSet: 'EPSG:3857',
                format: 'image/png',
                projection: projection,
                tileGrid: new ol.tilegrid.WMTS({
                  origin: ol.extent.getTopLeft(projectionExtent),
                  resolutions: resolutions,
                  matrixIds: matrixIds
                }),
                style: 'population',
                wrapX: true
            })
          })
        ],
        target: 'map',
        controls: ol.control.defaults({
          attributionOptions: {
            collapsible: false
          }
        }),
```

```
        view: new ol.View({
          center: [-11158582, 4813697],
          zoom: 5
        })
      });
      </script>
    </body>
</html>
```

You'll see our matrix IDs are being generated by a string that's just `EPSG:3857` and an incrementing integer corresponding to different zoom levels:

```
matrixIds[z] = 'EPSG:3857:'+z;
```

The resolutions are created by looking at the projection extent and dividing this by `256`:

```
var size = ol.extent.getWidth(projectionExtent) / 256;
```

This is a similar process to what we saw before with the gridsets. So, the OpenLayers web app is generating those values that correspond to the gridset for us so that, when it requests tiles, it matches up with the gridset.

And you can see origin resolution matrix IDs are all being defined here as part of the tile `grid` attribute, which is a part of the WMTS OpenLayer source object, as shown in the following code snippet:

```
tileGrid: new ol.tilegrid.WMTS({
    origin: ol.extent.getTopLeft(projectionExtent),
    resolutions: resolutions,
    matrixIds: matrixIds
```

Once all of this information is defined, our important endpoint here is to see that, the format is at our `geoserver` with the `gwc` section service and then `wmts` for `wmts` tile scheme endpoints as shown in the following code snippet:

```
url: 'http://localhost:8080/geoserver/gwc/service/wmts'
```

Once all these are defined, and it's matched up with the `map` through this `target` parameter, when we load our `map`, we will be getting tiles from that tile source.

In this section, you learned how to create an OpenLayers app using a GeoServer WMTS-style cache. In the next section, we'll view this tile-backed OL app in the browser, and you'll learn more about how tiles are accessed and created.

Using the tile cache

In the previous section, you learned how to create a tile-backed app using OpenLayers. In this section, we'll take a deeper dive into using the tile cache. We'll start by viewing the app we created in the last section in the browser. Next, we'll send a request for a tile generated by the app, but this time with the command-line utility `curl`. Finally, you'll learn about seeding and rendering the tile cache.

Here is the OpenLayers app that we created in the previous section:

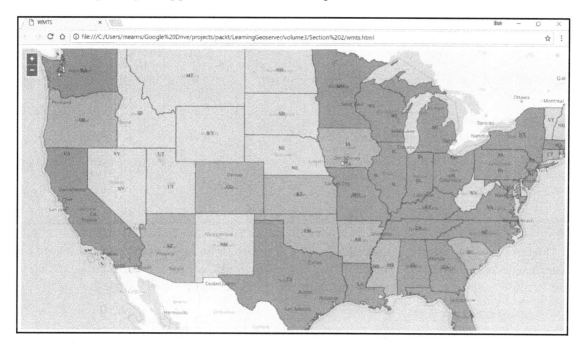

We'll use **Developer tools** from the browser again to see a URL for a particular tile, and then see what that looks like in `curl`. You can see we have tile requests here to OpenStreetMap, and also to our GeoServer instance:

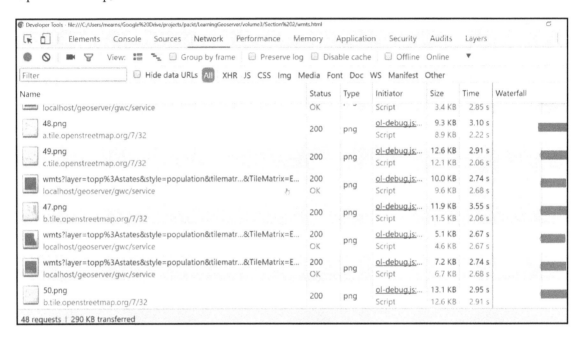

I'll just grab one of these link addresses, URLs, and then, moving over into the command line, if we already have `curl` installed, we can use the `curl` command with the `-v` switch for verbose output, so we can see information that will be useful for debugging:

```
C:\Users\mearns>curl -v
```

Then, paste the URL that we just copied, and of course this has all the mandatory WMTS parameters. And we get a response that includes some binary information, which will just appear as garbled text. You can see that this web cache result is a **HIT**, which means the tile was available in the tile cache, and GeoWebCache was able to respond directly with the pre-rendered image tile:

```
< HTTP/1.1 200 OK
< X-Frame-Options: SAMEORIGIN
< geowebcache-tile-index: [32, 78, 7]
< Content-Type: image/png
< geowebcache-cache-result: HIT
< geowebcache-tile-index: [32, 78, 7]
< geowebcache-tile-bounds: -10018754.17,4383204.949375,-9705668.1021875,4696291.0171874985
< geowebcache-gridset: EPSG:3857
< geowebcache-crs: EPSG:3857
< Last-Modified: Sat, 06 Jan 2018 16:41:36 GMT
< Content-Disposition: inline; filename=geoserver-dispatch.image
< Content-Length: 9794
< Server: Jetty(9.2.13.v20150730)
<
```

Now, let's see what this looks like when we miss a tile. In other words, this is what you would see if the tile was not pre-rendered. We can simulate this by deleting the entire tile cache directory for the layer that we wish to look at:

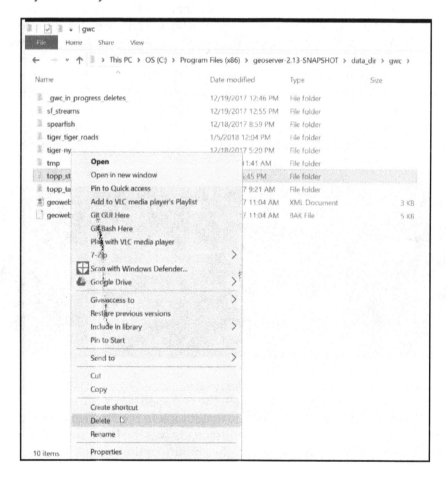

Running the same `curl` command, we'll see that this web cache result was a **MISS**:

```
< HTTP/1.1 200 OK
< X-Frame-Options: SAMEORIGIN
< geowebcache-tile-index: [32, 78, 7]
< Content-Type: image/png
< geowebcache-cache-result: MISS
< geowebcache-tile-index: [32, 78, 7]
< geowebcache-tile-bounds: -10018754.17,4383204.949375,-9705668.1021875,4696291.0171874985
< geowebcache-gridset: EPSG:3857
< geowebcache-crs: EPSG:3857
< Last-Modified: Sat, 06 Jan 2018 16:43:53 GMT
< Content-Disposition: inline; filename=geoserver-dispatch.image
< Content-Length: 9794
< Server: Jetty(9.2.13.v20150730)
<
```

Of course, if we run that command again, we'll see that this time it was a hit, because that tile was pre-rendered dynamically with GeoWebCache. Because we already went to request that tile once, GeoWebCache had to render that tile dynamically, and when we went back to request that tile again, the tile was available and so GeoWebCache returns a **HIT**. In other words, it did not have to render that tile on the fly this time. `curl` provides many useful pieces of information when doing tile requests. You can see the index of the tile itself, corresponding to the row and column in the gridset. You can also view the zoom level of the tile that's being requested, the bounding box, the gridset information, the `crs` information, and so on. This information is very useful if we're not getting our expected result, as shown in the following screenshot:

```
< HTTP/1.1 200 OK
< X-Frame-Options: SAMEORIGIN
< geowebcache-tile-index: [32, 78, 7]
< Content-Type: image/png
< geowebcache-cache-result: HIT
< geowebcache-tile-index: [32, 78, 7]
< geowebcache-tile-bounds: -10018754.17,4383204.949375,-9705668.1021875,4696291.0171874985
< geowebcache-gridset: EPSG:3857
< geowebcache-crs: EPSG:3857
< Last-Modified: Sat, 06 Jan 2018 16:43:53 GMT
< Content-Disposition: inline; filename=geoserver-dispatch.image
< Content-Length: 9794
< Server: Jetty(9.2.13.v20150730)
<
```

Another useful thing to look at if you have logging enabled or you're running GeoServer directly from the command line with the batch file or a shell script is the debugging information; this includes the amount of time that it took to render the file or reply with the tile that's being requested. You can see 48 milliseconds was the last return, and if we go up to the previous time that this was requested, the amount of time it took to reply is 2142 milliseconds.

So, there's a huge time saving here with this single tile as far as milliseconds go; of course, a millisecond is not a very long period of time, but when you're requesting an entire map with many tiles, these time savings will add up, and this is where we get the performance with tile caching.

Now, we're going to look at seeding a tile cache. Seeding means rendering all of the tiles and the tile cache, pre-rendering them. In this case, you won't need to ever dynamically render tiles on the fly because they'll be already rendered for you, and that will achieve the savings that we just discussed. The way to do this is to go to the **Actions** section and then **Seed/Truncate**; this is in the tile layers sidebar page, shown in the following screenshot:

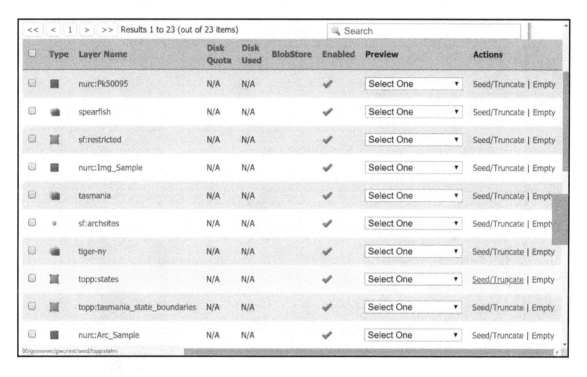

Another way to get to this is directly through GeoWebCache, which you can access by going to the GeoWebCache capabilities page. **Number of tasks to use** corresponds to the number of processes that will be used to render these tiles. This is bound by your CPU, so if you have multithreading, maybe a multi-core, multiprocessor system, you can use more tasks; if you have a less capable CPU on your system, you'll want to use fewer. I've had good results with **14**. In the past I wasn't really able to exceed four tasks, as I'd run out of memory.

That's another issue here: memory bounding. So, you'll want to select a number of tasks that matches the capabilities of your system. If we're seeding this for the first time, we can just use the **Seed - generate missing tiles** option. If we changed something on the backend in our data, perhaps we'll want to **Reseed - regenerate all titles** and **Truncate - remove titles.** This would just be the same as deleting, which you can also achieve just by deleting it on the filesystem. Select the gridset that's relevant, and we've defined the stop and start zooms already in this tile layer. Let's proceed everything at zoom five and six. The bounding box we can populate, and in particular if we know the area that needs to be receded; if we don't have that information, or if we just want the entire cache receded, we can just leave this bounding box blank. If you have any filter parameters on here, we'll need to fill those out, and if you're using more than one style, your screen will now look like this:

Create a new task:

Number of tasks to use:	14 ▾
Type of operation:	Seed - generate missing tiles ▾
Grid Set:	EPSG:3857 ▾
Format:	image/png ▾
Zoom start:	05 ▾
Zoom stop:	06 ▾
Modifiable Parameters:	STYLES: population ▾
	FORMAT_OPTIONS: dpi:72 ▾
Bounding box:	
	These are optional, approximate values are fine.
	Submit

After clicking on the **Submit** button, you can see we now have those 14 seed tasks running:

Id	Layer	Status	Type	Estimated # of tiles	Tiles completed	Time elapsed	Time remaining	Tasks	
2	topp:states	RUNNING	SEED	192	-1	Estimating...	Estimating...	(Task 2 of 10)	Kill Task
3	topp:states	RUNNING	SEED	192	-1	Estimating...	Estimating...	(Task 3 of 10)	Kill Task
4	topp:states	RUNNING	SEED	192	-1	Estimating...	Estimating...	(Task 4 of 10)	Kill Task
6	topp:states	RUNNING	SEED	192	-1	Estimating...	Estimating...	(Task 6 of 10)	Kill Task
7	topp:states	RUNNING	SEED	192	-1	Estimating...	Estimating...	(Task 7 of 10)	Kill Task
9	topp:states	RUNNING	SEED	192	-1	Estimating...	Estimating...	(Task 9 of 10)	Kill Task
10	topp:states	RUNNING	SEED	192	-1	Estimating...	Estimating...	(Task 10 of 10)	Kill Task
12	topp:states	RUNNING	SEED	192	-1	Estimating...	Estimating...	(Task 12 of 10)	Kill Task
13	topp:states	RUNNING	SEED	192	0	Estimating...	Estimating...	(Task 13 of 10)	Kill Task
14	topp:states	RUNNING	SEED	192	0	Estimating...	Estimating...	(Task 14 of 10)	Kill Task

Refresh list

As you're running through these, you can refresh this list. This ran through very quickly since we're at a high zoom. If you remember the example of the tile matrix, the number of tiles at lower zoom levels, in other words, lower scales, is going to be lower, so it takes much less time to render or seed those tiles. We can see the result of this on the filesystem if we want. Go to the `gwc` directory under the data directory, and you can see the top states directory, which corresponds to this layer, and you can see five through six are now seeded, so we have all tiles that will correspond to this bounding box:

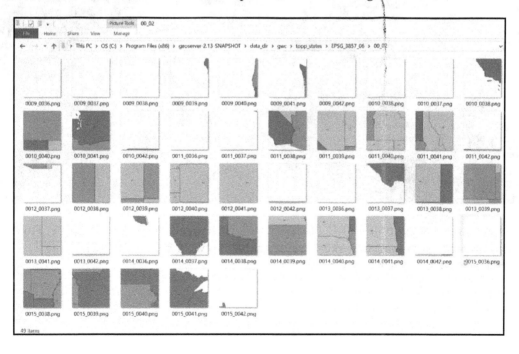

In this section, we took a closer look at using the tile cache and how tiles are created on the filesystem. In the next section, we'll explore some common difficulties with tile caching and how to resolve them.

Resolving problems with tile caching

In the previous section, you learned how to use the GeoServer tile cache. In this section, you'll learn about resolving issues that may arise with tile caching. First, you'll learn how to resolve the most common source of tile caching errors: parameters. Next, you'll learn about issues arising from rendering tiles. Finally, we'll review some other issues that may occur. The following example is showing a particular URL that includes parameters running through `curl`, which you saw in the previous section:

```
* timeout on name lookup is not supported
*   Trying ::1...
* Connected to localhost (::1) port 8080 (#0)
> GET /geoserver/gwc/service/wmts?layer=topp%3Astates&style=population&tilematrixset=EPSG%3A3857&Se
rvice=WMTS&Request=GetTile&Version=1.0.0&Format=image%2Fpng&TileMatrix=EPSG%3A3857%3A28&TileCol=12&
TileRow=25 HTTP/1.1
> Host: localhost:8080
> User-Agent: curl/7.45.0
> Accept: */*
>
< HTTP/1.1 400 Bad Request
< X-Frame-Options: SAMEORIGIN
< Content-Type: text/xml
< Content-Disposition: inline; filename=geoserver-dispatch.text
< Content-Length: 459
< Server: Jetty(9.2.13.v20150730)
<
<?xml version="1.0" encoding="UTF-8"?>
<ExceptionReport version="1.1.0" xmlns="http://www.opengis.net/ows/1.1"
    xmlns:xsi="http://www.w3.org/2001/XMLSchema-instance"
    xsi:schemaLocation="http://www.opengis.net/ows/1.1 http://geowebcache.org/schema/ows/1.1.0/owsExc
eptionReport.xsd">
    <Exception exceptionCode="InvalidParameterValue" locator="TILEMATRIX">
      <ExceptionText>Unknown TILEMATRIX EPSG:3857:28</ExceptionText>
    </Exception>
</ExceptionReport>
* Connection #0 to host localhost left intact

C:\Users\mearns>
```

I changed a parameter in this URL; the tile matrix is changed to be `28, 3857  :28`. This is not a tile matrix parameter that exists in the gridset, so we get an error back: `InvalidParameterValue`, related to `TILEMATRIX`.

So, to resolve an error of this sort, you would need to add `385728` to the gridset for this layer. The general area of parameter errors is very common when dealing with tile caches because the point of tile caches is to make dynamic things static, and that's really how tile caches achieve a better performance than anything that would be dynamically rendered. Whenever we introduce anything that will be dynamic into the request for a tile cache tile, the tile cache will want to either dynamically render that, which will hit their performance, or it's outside of the range of parameter values that's allowed. Another issue is that, sometimes, that parameter is not even a parameter the tile cache endpoint has. So, the example that we saw earlier was `dpi`, and in these cases we need to add that parameter to the filter parameters on the configuration for the tile cache endpoint, and generally we'll want to do that in the XML that's related to that tile cache layer.

Another set of issues that we often deal with in tile caching is issues relating to the rendering of the tiles themselves. This can arise when we have tiles layered on one another, as is the case with any kind of transparency. Those tiles are rendered independently of each other, and they don't necessarily understand what is going to be on another tile. Another case where we have this issue is with labeling. Labeling actually becomes pretty difficult in tile caching, although I will say my previous experience with labeling in older versions of GeoServer and GeoWebCahe was much more difficult, and we actually had some difficulty creating errors in GeoWebCahe and tile caching. In this version, it seems like a lot of these issues have been resolved. One issue that you may face in labeling is with larger areas; this example is from a WMS tile cache, and you can see we're getting more labels than what we would normally expect in these large areas:

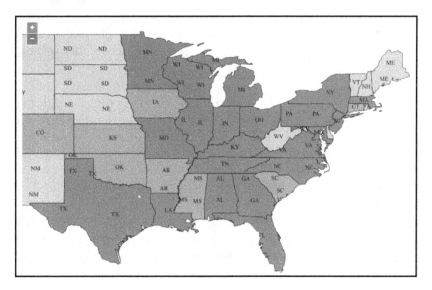

A way to resolve this would be to use centroids for these large areas. So the polygon is labeled at the center, rather than letting the algorithm figure out where to label it. The algorithm seems to want to label it in many places, and usually they're places where we would not expect these polygons to be labeled. To create centroids for these polygons, if you're working with a spatial database, you can easily create a view layer and just use a centroid function, then just label based on the view layer versus the layer itself. If we were working with a static data source, such as a shapefile, we would want to run that through our desktop GIS first and create a new static data source that is just centroids only.

Also, keep in mind that the SLDS, the styling for our particular layers, is going to be very important. And we may end up wanting to do additional styling, especially using the max scale, min scale denominator tag in your XML, your SLD style for your particular layer.

Now that we've gone over the two most common sources of issues with GeoWebCahe and GeoServer tile caching, here's a list of other issues that you may come across. These mostly have to do with performance, but can also just cause your tile cache not to work:

- **Don't provide additional styles**: I would recommend, as I mentioned earlier, not providing additional styles for your layers. In this way, you limit the amount of rendering that needs to be done. If you do provide additional styles, make sure that you seed those layers so that rendering is not being done on the fly, and that you recognize that additional storage and CPU will need to be provided for that rendering.
- **Disk and quota**: Disk is a limitation. Again, when providing additional options to your tile cache, you're going to be storing more tiles, and a way of limiting the hit on the disk is to use the disk quota available through GeoServer. In the sidebar, there's an option for limiting the number of tiles that will be stored on the disk.
- **Java**: This seems to be less of an issue in more recent versions of GeoServer, which is a good thing of course. I ran into this many times in older versions of GeoServer. A way of getting around this if you do run into memory errors, GeoServer will just crash, and if you're running into those kinds of errors, you can put quotas and limits on tile caching. You can also expand the amount of memory that is available to Java, and that's done in the startup scripts, the shell script, or the batch script that you're using, whether you're using Linux or Windows.
- **Try nightly builds**: If you're running into a strange error, which I have seen even more recently, you might just want to try a nightly build. There is active development going on with tile caching and so sometimes those are fixed in these more recent builds, and I had that experience when I was writing this book.

In this section, you learned all about tile caches, including the basics of how tile caches and various schemes work, GeoServer settings for tile caches, integrating tile caching with an OpenLayers app, and examining a tile cache in the browser and from the command line with `curl`. In this section, you also learned how to resolve issues with tile caching.

Summary

This concludes the chapter on using tile caching to make your app faster. This was a challenging chapter, so give yourself a pat on the back for getting through it. In this chapter, we used tiles, pre-rendered portions of a map, to allow your app to load more rapidly. I covered the material you need to know to get your tile cache up and running, and fix any issues that may occur. In the next chapter, you'll learn how to enable additional GeoServer capabilities and tune your GeoServer instance for even better performance.

3
Optimizing GeoServer

In this chapter, you'll learn how to improve the reliability and speed of GeoServer by identifying instances where potential client demand outstrips hardware resources. First, you'll learn about bottlenecks and how they can be resolved. You'll then learn how to optimize vector and raster data stores. Finally, we'll cover the basics of clustered GeoServer deployment.

In this chapter, we will cover the following topics:

- Resolving bottlenecks
- Optimizing vector data stores
- Optimizing raster data stores
- Clustered deployment

Resolving bottlenecks

Let's get started with resolving bottlenecks. In this section, you'll learn why bottlenecks occur and how to go about resolving them. We'll start by exploring examples of bottlenecks and then learn how tuning can be used to resolve them. Finally, you'll learn methods of testing and benchmarking to use in tandem with tuning to detect bottlenecks iteratively.

Bottlenecks occur when resources don't match demand. On a hardware level, this can involve networking. If you have multiple servers, for example, a bottleneck can occur when the internal network is exceeded by the demand on the network between the servers, or if the network path from your servers to the client is exceeded. A bottleneck can also occur due to storage limits, which can be easily exceeded by large geospatial files. The disk's speed, memory, and capacity are all other causes of bottlenecks; GeoServer is a highly demanding application in terms of memory as well as processor capacity.

On a software level, GeoServer does contain some bugs that are important to be aware of. As you'll see later on in this chapter, it is possible to detect these bugs when the server meets its limits. These limits themselves could be due to bugs, but they more often come from hardware limitations or from other software limitations. In Java, for example, we have the JVM, as well as the servlet container or the application server that you might use. We need to make sure the GeoServer settings are configured appropriately, particularly with regard to rendering and data connections. This is because rendering on the browser instead of the server can be highly demanding and can cause bottlenecks. The reason to be aware of these different areas that may cause bottlenecks is that we can hone in on these using benchmarking logs and detect where the bottleneck is occurring.

In the following two global tuning examples, you'll learn how to resolve bottlenecks that occur throughout the GeoServer instance using tuning. The JVM settings shown in the following screenshot are a good example of the dilemma we deal with in tuning. These settings can also be found in the documentation (`http://suite.opengeo.org/opengeo-docs/geoserver/production/container.html`):

Optimize your JVM

Set the following performance settings in the Java virtual machine (JVM) for your container. These settings are not specific to any container.

Option	Description
Xms128m	By starting with a larger heap GeoServer will not need to pause and
Xmx756M	Defines an upper limit on how much heap memory Java will request
XX:SoftRefLRUPolicyMSPerMB=36000	Increases the lifetime of "soft references" in GeoServer. GeoServer u
XX:+UseParallelGC	The default garbage collector, **pauses the application while using**
XX:+UseParNewGC	Enables use of the concurrent mark sweep (CMS) garbage collector t
XX:+UseG1GC	Enables use of the Garbage First Garbage Collector (G1) using ba

If the resource allocation is too high, it will cause an out-of-memory error and GeoServer will crash. If we set our parameters too low, although this may result in greater stability on the server, it can cause requests to drop or slow down.

The easiest way to alter the JVM settings shown in the preceding screenshot is to alter the JAVA_OPTS environment variable. In the following screenshot, you can see an echo of the JAVA_OPTS environment variable:

```
Command Prompt
Microsoft Windows [Version 10.0.16299.192]
(c) 2017 Microsoft Corporation. All rights reserved.

C:\Users\mearns>echo %JAVA_OPTS%
-Xms128m -Xmx756M -XX:SoftRefLRUPolicyMSPerMB=36000

C:\Users\mearns>setx JAVA_OPTS "%JAVA_OPTS% --XX:+UseParNewGC"
```

These may be some of the settings that we already set. By default, you probably won't see anything when you echo the environment variable. There are, however, some parameters. One is related to the minimum amount of memory (Xms128m) used by GeoServer or the JVM when you start them up. We can bump this up to add memory so that GeoServer isn't trying to allocate this as it runs into higher demand for resources. We can also set a maximum amount of memory (Xmx756M) that the JVM can grab, which can prevent an out-of-memory error. The SoftRefPolicy parameter is related to the amount of memory that we give to the reference, which is related to the cache. If we bump it up, we'll have more information stored in the cache, which of course enhances performance. The downside, however, is that we might run into an out-of-memory error again if it takes up too much memory in the cache.

The command in Windows for setting environment variables is setx. The following is an example of how to add a new parameter to the JAVA_OPTS environment variable. The UseParNewGC parameter involves garbage collection:

```
C:\Users\mearns>setx JAVA_OPTS "%JAVA_OPTS% --XX:+UseParNewGC"
```

There are a number of garbage collection options that you can use with the JVM. This one in particular involves multiple threads for doing garbage collection, concurrent with other requests that may be going on in your GeoServer instance. The benefit here is that GeoServer doesn't need to pause to carry out this action, and garbage collection of course frees up memory.

The control flow extension limits the number of concurrent operations, such as requests and threads. Control flow offers a similar kind of tuning challenge; hiring a resource allocation such as concurrent requests to provide greater memory will give us an out-of-memory crash, but too little memory will result in greater latency.

In addition to the usual extension installation process, the control flow extension requires a manually created `controlflow.properties` file under the `data_dir` directory. The content for this can be found on the control flow reference page on the GeoServer site (`http://docs.geoserver.org/stable/en/user/extensions/controlflow/index.html`), or you can use the sample that we've provided with the book:

```
# if a request waits in queue for more than 60 seconds it's not worth
executing,
# the client will likely have given up by then
 timeout=60

# don't allow a single user to perform more than 6 requests in parallel
# (6 being the Firefox default concurrency level at the time of writing)
 user=6

# don't allow the execution of more than 100 requests total in parallel
 ows.global=100

# don't allow more than 10 GetMap in parallel
 ows.wms.getmap=10

# don't allow more than 4 outputs with Excel output as it's memory bound
 ows.wfs.getfeature.application/msexcel=4

# don't allow the execution of more than 16 tile requests in parallel
 # (assuming a server with 2 cores, GWC empirical tests show that
throughput
 # peaks up at 4 x number of cores. Adjust as appropriate to your system)
 ows.gwc=16
```

You can see that this deals with various factors including client timeouts, number of `GetMap` requests that can be made in parallel or concurrently, and the maximum concurrent number of Excel outputs that can be requested, as shown earlier. Moreover, you can hone in on operations that may require an unusual amount of resources from your server, which allows for a fairer use of resources across your entire user base.

To test that bottlenecks have been resolved, or to detect the possible presence of bottlenecks, it is important to use performance monitoring and logging tools. The comparison of performance indicators at different points in time is called benchmarking. It is important to benchmark before and after changes to hardware or software configurations to confirm that the bottleneck has been resolved.

You'll use some tools we've worked on before, such as `GeoServer` logs, which you can see here:

```
geoserver.log
2018-01-05 12:01:28,628 INFO [geoserver.filters] - 0:0:0:0:0:0:0:1 "GET
/geoserver/gwc/service/wmts?layer=tiger%3Atiger_roads&style=tiger_road_nosca
le&tilematrixset=EPSG%3A900913&Service=WMTS&Request=GetTile&Version=1.0.0&Fo
rmat=image%2Fpng&TileMatrix=EPSG%3A900913%3A11&TileCol=602&TileRow=769"
"Mozilla/5.0 (Windows NT 10.0; Win64; x64) AppleWebKit/537.36 (KHTML, like
Gecko) Chrome/63.0.3239.84 Safari/537.36"
"http://localhost:8080/geoserver/gwc/demo/tiger:tiger_roads?gridSet=EPSG:900
913&format=image/png" ""
2018-01-05 12:01:28,627 INFO [geoserver.filters] - 0:0:0:0:0:0:0:1 "GET
/geoserver/gwc/service/wmts?layer=tiger%3Atiger_roads&style=tiger_road_nosca
le&tilematrixset=EPSG%3A900913&Service=WMTS&Request=GetTile&Version=1.0.0&Fo
rmat=image%2Fpng&TileMatrix=EPSG%3A900913%3A11&TileCol=600&TileRow=769"
"Mozilla/5.0 (Windows NT 10.0; Win64; x64) AppleWebKit/537.36 (KHTML, like
Gecko) Chrome/63.0.3239.84 Safari/537.36"
"http://localhost:8080/geoserver/gwc/demo/tiger:tiger_roads?gridSet=EPSG:900
913&format=image/png" ""
2018-01-05 12:01:28,627 INFO [geoserver.filters] - 0:0:0:0:0:0:0:1 "GET
/geoserver/gwc/service/wmts?layer=tiger%3Atiger_roads&style=tiger_road_nosca
le&tilematrixset=EPSG%3A900913&Service=WMTS&Request=GetTile&Version=1.0.0&Fo
rmat=image%2Fpng&TileMatrix=EPSG%3A900913%3A11&TileCol=602&TileRow=768"
```

You'll note that, when a resource is requested, you'll be able to see how long it takes for GeoServer to return the request. Browser developer tools, especially the network tab, allow you to see system performance logs and how long it takes to process a request. A performance-monitoring program to produce logs and to see the performance of different resources on your server can be really useful. One particular program that you would want to check out if you're on Windows is Performance Monitor. This produces a very useful graphical output that you can use to see where bottlenecks may be occurring:

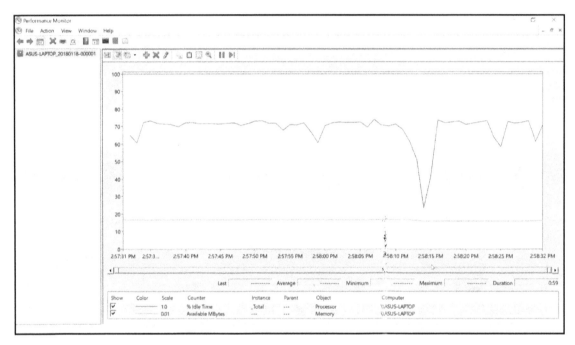

For example, if you're seeing a gradually increasing output or throughput on your network device and then it hits a maximum and drops off sharply, as shown in the preceding screenshot, you'll know that a bottleneck has occurred. A similar type of visual representation that you'll see for other hardware resources. JMeter is a great option for doing GeoServer benchmarking, because, unlike simple system logs, JMeter allows us to simulate multiple requests on the server. You can always use JMeter in tandem with other locking options.

To produce this sort of test on JMeter, you just create a thread group that includes the option to use multiple threads. You then create a loop, which runs a request or some sort of operation in a loop. In this way, you can test what it would be like to get these multiple concurrent operations on your server.

It's best to do this with a remote JMeter instance so the resources that it takes to run JMeter are not interfering with what you're seeing for your server performance. This is even more useful when you can work with clusters. JMeter is a very sophisticated program because you can also include various other features, such as random requests.

In this section, we've looked at what bottlenecks are and how to go about detecting and resolving them with a few examples. In the next section, we'll look at optimizing vector data stores and ways of resolving the common sources of bottlenecks.

Optimizing vector data stores

In the previous section, you learned how to identify bottlenecks and resolve them by tuning. In this section, you will learn how to configure vector stores for maximum performance, thereby eliminating a possible source of bottlenecks. First, we'll look at data storage, a factor that impacts vector and raster data performance. Next, we'll learn about the best formats for our vector data, based on their intended use. Finally, we'll learn about a topic that's particularly important to vector performance – indexing.

Choosing a data storage location is a relatively simple topic that has a big impact on performance if it is overlooked. Since the capacity is less of a concern with vector data than raster data, speed is the most important factor here. Whenever practical, opt for data that is stored locally on the GeoServer server, paying special attention to the nominal read/write speed of the drive or drive interface on which you intend to store the data. If you using a hard drive, make sure you note the RPM, which relates to latency. To ensure a drive is performing at its rated speed, or to incorporate the impact of latency, you can benchmark drives relative to one another by using system logging or performance monitoring software, as seen in the previous section.

GeoServer makes it easy to update the software backup and migrate servers by having a `data_dir` directory that is independent of the rest of the software. You should move it to an external location outside of the GeoServer directory, as shown in the following screenshot:

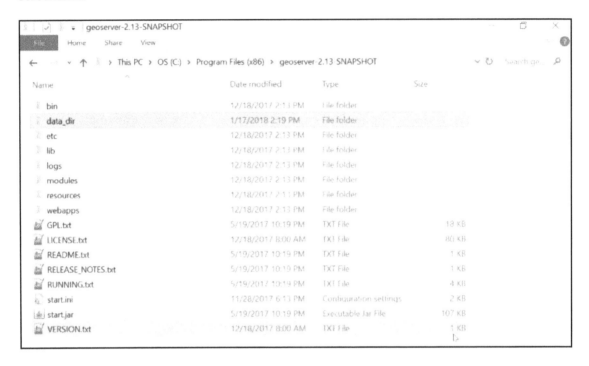

The easiest way to do this is by changing the environment variable, `GEOSERVER_DATA_DIR`. This is the GeoServer data directory location. You can do this on Windows through the system properties, or use the commands you saw in the previous section. You need to echo out the environment variable with percentage marks around it to see what it is, and then use `setx` to set a new environment variable with a new data directory location, as shown in the following screenshot:

If the data is being consumed by multiple GeoServer nodes, as is the case in a cluster, a storage area network (SAN) should be used. If using a SAN or shared disk, the environment variable GEOSERVER_REQUIRE_FILE should be used. This variable will be set to a file that's on the filesystem under the data_dir directory:

```
setx GEOSERVER_REQUIRE_FILE "F:\somedir\data_dir\somefile"
```

This tells the active GeoServer node that the data_dir directory is active and will not fail; otherwise, it will exit gracefully. Multiple instances of the GEOSERVER_REQUIRE_FILE can also be used if there is a possibility that one of the files might not be available. In such a case, you do not want the GeoServer instance to exit gracefully, so it will continue to run.

On a database, the data will always be stored outside the GeoServer data or program directories. What we should be paying attention to is the connection pool. There are a few parameters related to the connection pool and to how connections are made between GeoServer and the database. Examples of these parameters include **max connections** and **min connections**, as shown in the following screenshot:

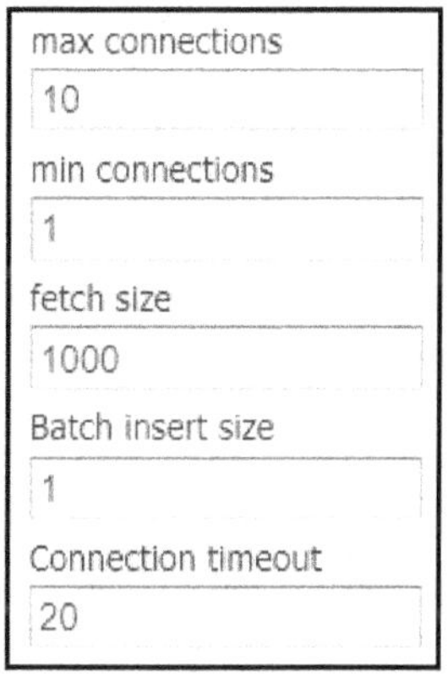

max connections
10

min connections
1

fetch size
1000

Batch insert size
1

Connection timeout
20

If **min connections** is set above 1, a connection pool is created. This creates multiple concurrent connections to the database so that connections will not need to be made on the fly. This is especially important if the database is not local. That way, GeoServer does not need to establish a connection every time data is requested from the database, since these connections will already be maintained in a pool.

Vector data formats

When looking at vector data formats, the first thing to recognize is that database stores are generally the best. Among database servers, PostGIS is the most suitable for GeoServer. PostGIS is well supported by the GeoServer community and has additional capabilities. It's also just a very good spatial database in general. You will definitely want to use a spatial database if you're working with larger datasets, because you will sometimes only be reading a portion of those datasets or filtering dynamic requests against those datasets. You will learn more about this in a moment, when we talk about indexing. You will also definitely want to use a database if you're working with WFS-T, which is writing against GeoServer. Otherwise, the shapefile is actually a pretty good format and a lot of you are probably already using it. The benefit of shapefile, apart from its familiarity, is that it's already in a compressed format and you won't need to worry about rendering or pulling the data out on the fly spatially, since this will be pretty fast. You should avoid WFS and text-based formats whenever they are not needed. This includes GeoJSON, GML, and KML, among others.

Indexes in data stores are similar to indexes in a book. Data is sorted ahead of time into an index, which supports fast lookup via a rule or a set of rules. In a book, indexes use alphabetized lists of important topics and the page numbers where the topics can be found. In the case of GIS data stores, clever algorithms are used to index, the feature spatial extents and attributes. Database stores offer the best options when it comes to indexing. As you can see here, in this example of creating an index with PostGIS, you have the option to select a particular table. We then have the following options for the type of index to be created:

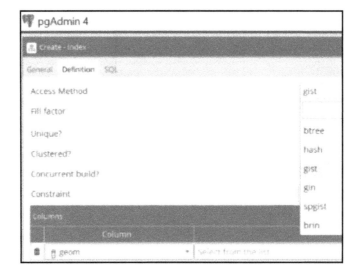

`gist` is the index that you'll be using if you're doing spatial indexing, which is indexing around spatial extents. This is useful if you're doing any kind of spatial operation. This will speed up the reading of the data on the database. Other indexes are used for attribute indexing – if you're filtering against an attribute, for example – or if you're doing any kind of subsetting against attributes connected to spatial features.

Shapefiles can also be indexed, but only spatially. You can do this by checking the **Create spatial index if missing/outdated** checkbox on the data store, as shown in the following screenshot:

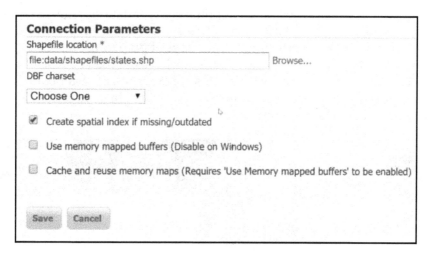

Creating a spatial index in this way will produce a QIX file or QIS file among the files that are connected to the shapefile, as shown in the following screenshot:

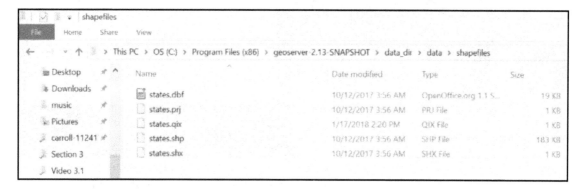

You can create this using GDAL, but this is not recommended. Using GeoServer is supposed to be faster. If you have already created one via GDAL, you might want to consider deleting that and creating one via GeoServer using the method just outlined. In this section, we've learned how to optimize vector data. In the next section, we will learn about the additional challenges of optimizing raster data.

Optimizing raster data stores

In the previous section, you learned how to optimize vector data storage. In this section, you'll learn how to do so for raster data. First, we'll cover a few concepts regarding the processing of raster data. We'll learn about some ways of preparing data using the command-line utility GDAL. Finally, we'll take a quick look at the WCS OGC service offered by GeoServer. Imagery and analysis grids are examples of raster data. Raster processing is a set of operations that are used to make grid data more useful. The following diagram contains a high-resolution photograph of a cat on the left-hand side:

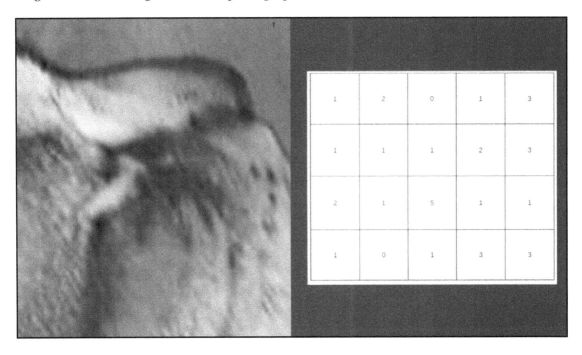

We have to zoom in pretty close before we can start to see pixels. Resolution refers to the number of pixels or grid cells per square inch, or some other unit of linear measure. Each of these pixel cells contains a number, which is then interpreted as colors by the GIS program or by GeoServer.

Sampling is the conversion of one particular resolution into an image of a different resolution. The number of grid cells per inch is changed. Compression is similar to sampling. Sampling uses a set of algorithms that determine what the new cell value will be, based on the old cell value or the values of the neighboring cells. Compression uses a more sophisticated algorithm to determine what values the new cells will have. The following image is compressed to 10 times smaller than the size of the original:

By using compression, the image still retains some fidelity to the original, despite the high resolution reduction and the much smaller size. Compression is useful if you have a bottleneck related to disk size or network data transfer. However, since compression does require some additional disk processing with GeoServer, it effectively slows down the data access rate from your drive. The `gdal_translate` command is a utility which is part of the GDAL package. It is very useful for resampling images. GeoTIFF is the format that is most widely recommended for geospatial raster data in GeoServer. GeoTIFF is a non-compressed image format.

The following image is a single tile from a larger vector data image:

GeoServer, or any tiling software, uses the same process, more or less, to create tiles from the original data. You've seen tiling in Chapter 2, *Speed Up Your App with Tile Caching*. Tiling improves performance since tiles speed up the viewing of data on the browser. Rendering is done ahead of time on the server, rather than on the client machine. Mosaicing is a similar concept to tiling. If you have imagery that's captured in tiles, such as aerial imagery, that you want to put together as a single scene that shows an entire state, you would use a mosaic to stitch these separate images together. There is a mosaic plugin for GeoServer that allows you to do this on the fly, but it does have an impact on performance. We will not be covering that process here. GeoTIFF supports internal tiling, so we can get some of these benefits directly from the image file format.

An overview is a sampled or down sampled, lower-resolution version of the original image, similar to if you had zoomed out from the image. The result is a lower-resolution image that GeoServer knows to load if the user is zoomed out, resulting in better performance, as shown in the following diagram:

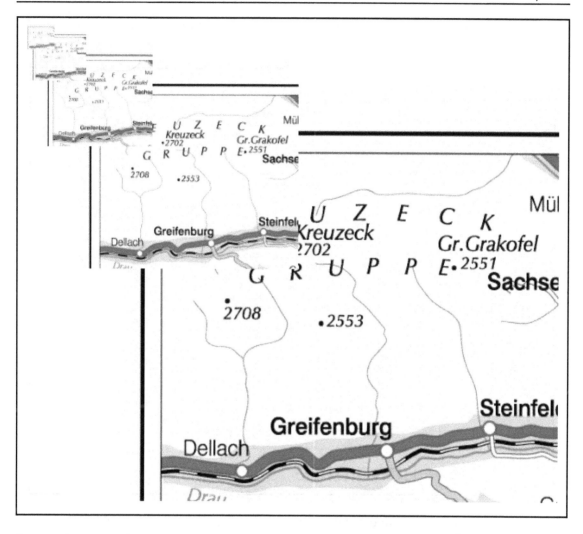

Pyramids are similar to overviews, but are used with mosaic datasets that encompass many tiles. GeoTIFF also supports internal overviews.

A multidimensional raster dataset includes additional dimensions, such as time and elevation, as shown in the following diagram:

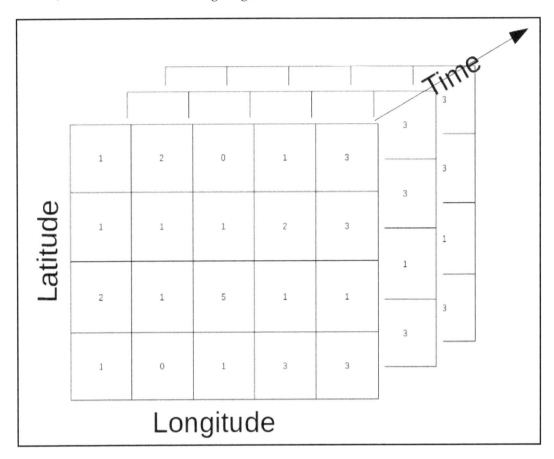

NetCDF is a common format and can be used with GeoServer via the NetCDF reader extension.

Let's now use the GDAL raster package to enhance a raster image for production. The first command that will run is the `gdalinfo Pk50095.tif` command, which gives us some very useful information about our image, as shown in the following screenshot. A lot of this information is not easily available elsewhere:

```
Command Prompt                                                              —   □   ×
C:\Users\mearns\Google Drive\projects\packt\LearningGeoserver\volume3\Section 3\Video 3.3\image>gda
linfo Pk50095.tif
Driver: GTiff/GeoTIFF
Files: Pk50095.tif
       Pk50095.tfw
Size is 545, 490
Coordinate System is `'
Origin = (347649.930868591070000,5196961.352859255900000)
Pixel Size = (42.341367999999903,-42.341367999999903)
Metadata:
  TIFFTAG_RESOLUTIONUNIT=2 (pixels/inch)
  TIFFTAG_SOFTWARE=IrfanView
  TIFFTAG_XRESOLUTION=200
  TIFFTAG_YRESOLUTION=200
Image Structure Metadata:
  INTERLEAVE=PIXEL
Corner Coordinates:
Upper Left  (   347649.931, 5196961.353)
Lower Left  (   347649.931, 5176214.083)
Upper Right (   370725.976, 5196961.353)
Lower Right (   370725.976, 5176214.083)
Center      (   359187.954, 5186587.718)
Band 1 Block=545x5 Type=Byte, ColorInterp=Red
Band 2 Block=545x5 Type=Byte, ColorInterp=Green
Band 3 Block=545x5 Type=Byte, ColorInterp=Blue

C:\Users\mearns\Google Drive\projects\packt\LearningGeoserver\volume3\Section 3\Video 3.3\image>gda
l_translate -of GTiff -co "TILED=YES" Pk50095.tif Pk50095_tiled.tif
Input file size is 545, 490
```

This tells us the size of the image, information about the coordinate system, where the coordinates begin, information about resolution and bands, and, importantly, the block size of the image. The image is composed of strips or stripes, as shown in the previous screenshot, and we want something more like square tiles, which are efficiently loaded in GeoServer.

To get tiles in your new TIFF, we'll use the gdal_translate utility with a TILED=YES parameter, as shown in the following screenshot:

```
C:\Users\mearns\Google Drive\projects\packt\LearningGeoserver\volume3\Section 3\Video 3.3\image>gda
l_translate -of GTiff -co "TILED=YES" Pk50095.tif Pk50095_tiled.tif
Input file size is 545, 490
0...10...20...30...40...50...60...70...80...90...100 - done.
```

This will produce an internally tiled file. We know that this file is now tiled because, if we look at the block size, using the gdalinfo Pk50095_tiled.tif command, it is now 256 x 256, as shown in the following screenshot. This is a good standard tile size:

```
Band 1 Block=256x256 Type=Byte, ColorInterp=Red
Band 2 Block=256x256 Type=Byte, ColorInterp=Green
Band 3 Block=256x256 Type=Byte, ColorInterp=Blue
```

Finally, gdaladdo adds internal overviews to the tile. We're adding overviews at the 2, 4, 8, and 16 zoom levels, as shown in the following snippet:

```
$ gdaladdo -ro -r average Pk50095_tiled.tif 2 4 8 16

0...10...20...30...40...50...60...70...80...90...100 - done.
```

This refers to the size of the original image being reduced by 50%, 25%, and so on. In both cases, these utilities, gdal_translate and gdaladdo, are adding internal tiles and internal overviews to the files, so we won't necessarily see a separate file as a result. We don't see the tile files we might be used to seeing, or an overview file external to the TIFF itself, but we can use the gdalinfo utility to check that the overviews were created. Now, we can see the overviews for each of these bands, as shown in the following screenshot:

```
Band 1 Block=256x256 Type=Byte, ColorInterp=Red
    Overviews: 273x245, 137x123, 69x62, 35x31
Band 2 Block=256x256 Type=Byte, ColorInterp=Green
    Overviews: 273x245, 137x123, 69x62, 35x31
Band 3 Block=256x256 Type=Byte, ColorInterp=Blue
    Overviews: 273x245, 137x123, 69x62, 35x31
```

WCS

WCS is an OGC web service while GeoServer is used for imagery. You could use WMS, but WCS has more capabilities. It is sometimes referred to as the WFS of raster imagery because it exposes more information about the raster data that we're working with to the end user. Along with the metadata, it also allows you to subset the raster dataset, which is useful if you're working with a large dataset. There are also more formats available as outputs.

The request builder is built into GeoServer, and allows us to construct a request and see the result, similar to the GeoServer demo request builder. This is useful for generating requests and response XMLs to mock up a WCS web client.

In the following particular subset, I took the original values and modified them to create a smaller extent as the subset:

When you click **Get Coverage**, this request will produce a downloadable file.

The following screenshot is the result of that operation:

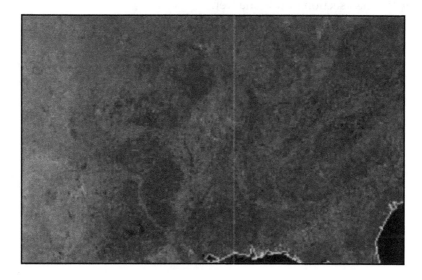

This is a subset of a national raster dataset. In this section, we have learned how to optimize raster data for GeoServer. In the next section, we'll learn about server clustering and how to use it to improve performance and reliability.

Clustered deployment

In the previous section, we looked at how to best deal with large raster datasets. In this section, we'll explore clustered server deployment to get the best GeoServer user experience from our hardware resources. We'll learn about concurrency and clusters and why these concepts are so important. We'll also look at how to use vertical cluster deployment for higher performance. Finally, we'll learn about GeoServer horizontal clusters for high availability.

Concurrency refers to tasks being run simultaneously on different parts of a computer or cluster hardware. GeoServer allows us to use concurrency in a few different ways. It can be deployed in a clustered configuration on one or more nodes. A cluster refers to a configuration of multiple computing resources, which are either virtual machines or physical computers (nodes). This means that tasks can be run more efficiently using concurrency. We'll be covering this later in this section.

In the previous section, we learned about the use of multiple threads to run GWC tile seeding. This allows multiple tiles to be created simultaneously, dramatically decreasing the time it takes to seed a cache. The JVM is multithreaded and so it can process concurrent requests from web users. This is managed by the control flow extension that you saw in the *Resolving bottlenecks* section of this chapter.

Multithreaded software achieves concurrency by taking advantage of multiple compute threads within the compute resources of a single node. It is also possible to cluster the database that GeoServer uses for data stores. Current versions of Postgres offer some of the benefits of clustering, such as streaming replication to a hot standby read-only database. There is also a bevy of extensions that focus on aspects of high availability or load balancing. I won't go into detail about Postgres clustering here. You can read more about it at `https://www.postgresql.org/docs/9.1/static/sql-cluster.html`.

Clusters offer concurrency. The question is, however, how does this specifically benefit the end user? Different configurations offer different benefits, but adding additional hardware and complex architecture comes at a cost. If a single VM or a hardware node fails, or when services such as software updates must be stopped for any reason, requests can still be routed to an active replacement with little disruption to the end user. This is called high availability. If more hardware is used, requests can be processed more quickly, which is a performance benefit. As a service becomes more heavily used, the load on the hardware increases. The load can be distributed across an expandable pool of hardware resources. This is known as scalability.

In a vertical cluster, a single node with extensive hardware uses multiple GeoServer instances. The location of the data directory is very important with GeoServer clusters. The data for all instances should remain in sync. This is even more important with the GeoServer XML configuration files, all of which are stored in the data directory. If configuration files differ between cluster instances, the entire cluster is likely to become unstable. This is why a shared data directory location is important. In a vertical cluster, the location of the data directory is easily resolved, as both virtual machines will have local access to the disk where the data directory is stored.

Another important issue is load balancing. Load balancing is usually performed by standalone software such as HAproxy. Alternatively, it can be performed by HTTP software such as Apache HTTPD, with mod proxy or mod proxy aJP, which are formally known as mod_JK modules. An example of a vertical cluster is a single node with two JVMs running two Tomcat instances, each with its own private port, as depicted in the following diagram:

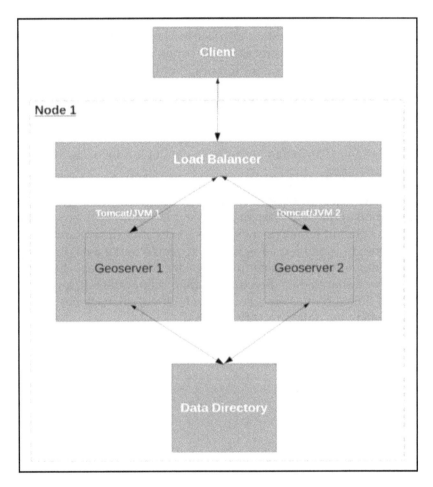

A load balancer, such as HAproxy, is installed to listen for all requests at the public server address and port. In the simplest case, HAproxy will alternate the Tomcat instances to which the request is sent in round-robin fashion, so user requests are distributed across available hardware more efficiently than with a single JVM.

A horizontal cluster is configured across multiple nodes. The most important benefit of horizontal clustering is high availability, as it avoids single points of failure. In other words, if a server goes down for any reason, whether planned or unplanned, the GeoServer cluster will still respond to the requests of the end user. In a horizontal cluster, the application server may also be used for load balancing. Load balancers can be configured on one or more external nodes or on the GeoServer nodes themselves. If fewer than two nodes are used for the load balancer, the node hosting the load balancer becomes a single point of failure, thereby compromising high availability. In one configuration, depicted in the following diagram, the data directory is in a shared location:

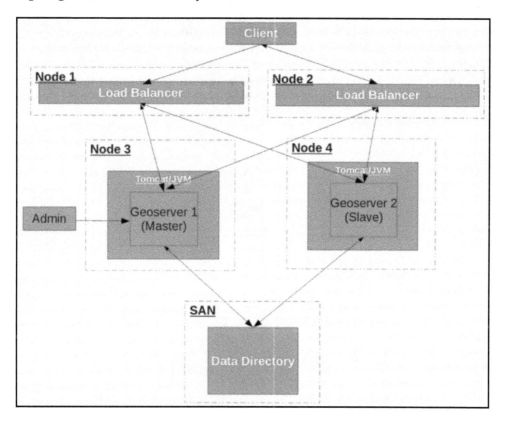

Ideally, this is a fast storage area network drive. Any administrative tasks that change the configuration files will be done on the master. The slave is set to read-only and the configuration can be loaded when appropriate. The SAN becomes the single point of failure.

In another configuration, the data directories are on separate nodes. This is the most highly available configuration since there is no single point of failure. The data directories must be synced using some external software such as rsync. Each GeoServer instance is a master and slave. They communicate administrative configuration tasks using a broker, the GeoServer active clustering extension, which uses the Java messaging service to communicate.

Summary

This was another challenging chapter; great job for sticking with it. In this chapter, you learned how to improve GeoServer's reliability and speed by identifying instances where potential client demand outstrips hardware resources. You learned about bottlenecks and how they can be resolved. You also learned how to optimize vector and raster data stores. Finally, we covered the basics of clustered GeoServer deployment. In the next chapter, you will learn how to secure your GeoServer instance and web application.

4
Secure Authentication

In this chapter, you will learn how to use HTTPS to provide secure login from an OpenLayers app to your GeoServer-based services. You'll also learn about the building blocks of our proxy platform, Apache HTTPD, and Tomcat, and how to configure them. Next, we'll cover HTTPS and TLS, and you'll learn how to generate certificates to enable secure requests. You'll learn about GeoServer's user-access permissions that permit authorized users to access the server. Finally, we'll test out a secure login with curl and an OpenLayers web app.

In this chapter, we will cover the following topics:

- Configuring the proxy
- HTTPS with TLS and certificates
- GeoServer authentication
- Secure login from OL

Configuring the proxy

Let's get started by configuring the proxy environment. A proxy server is widely used to provide a secure means of authentication to the GeoServer instance on Tomcat. In this section, you'll learn about the main components of the example proxy environment; Apache HTTPD and Tomcat. You'll learn how to configure these to provide secure authentication. Finally, you'll learn to use the hosts file to facilitate the TLS process.

The Apache HTTP server—HTTPD for short—is the most popular web server in use. Putting HTTPD in front of Tomcat is useful since we can potentially leverage a larger assortment of modules and their capabilities, and take advantage of its load balancing functionalities and the larger body of documentation.

After Apache is correctly installed and started, the default page will be available at the localhost. I used one of the packages that is available out there on the web. The following one is for Windows I used it to install Apache rather than build it myself:

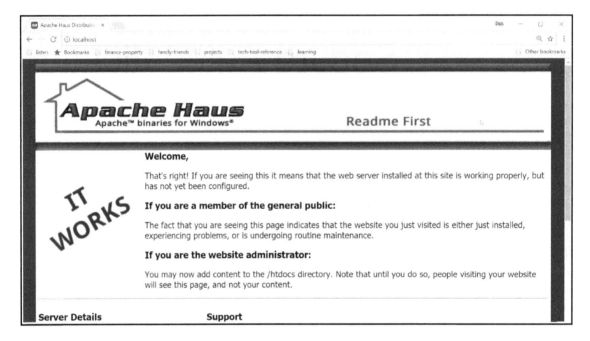

The use of such packages has the benefit of having many prebuilt modules at your disposal, and also comes with the open SSL executables that we'll use to produce the certificate in the next section. To configure Apache, check out the `httpd.conf` file under the `conf` directory, where Apache is installed. The `httpd.conf` file is well commented and relatively easy to read. The configuration syntax is usually referred to as directives.

The following section of the `httpd.conf` file shows the modules that are loaded with HTTPD:

```
# Example:
# LoadModule foo_module modules/mod_foo.so
#
#LoadModule access_compat_module modules/mod_access_compat.so
LoadModule actions_module modules/mod_actions.so
LoadModule alias_module modules/mod_alias.so
LoadModule allowmethods_module modules/mod_allowmethods.so
LoadModule asis_module modules/mod_asis.so
LoadModule auth_basic_module modules/mod_auth_basic.so
#LoadModule auth_digest_module modules/mod_auth_digest.so
#LoadModule auth_form_module modules/mod_auth_form.so
```

```
#LoadModule authn_anon_module modules/mod_authn_anon.so
LoadModule authn_core_module modules/mod_authn_core.so
#LoadModule authn_dbd_module modules/mod_authn_dbd.so
#LoadModule authn_dbm_module modules/mod_authn_dbm.so
LoadModule authn_file_module modules/mod_authn_file.so
#LoadModule authn_socache_module modules/mod_authn_socache.so
#LoadModule authnz_fcgi_module modules/mod_authnz_fcgi.so
#LoadModule authnz_ldap_module modules/mod_authnz_ldap.so
LoadModule authz_core_module modules/mod_authz_core.so
#LoadModule authz_dbd_module modules/mod_authz_dbd.so
#LoadModule authz_dbm_module modules/mod_authz_dbm.so
LoadModule authz_groupfile_module modules/mod_authz_groupfile.so
LoadModule authz_host_module modules/mod_authz_host.so
#LoadModule authz_owner_module modules/mod_authz_owner.so
LoadModule authz_user_module modules/mod_authz_user.so
LoadModule autoindex_module modules/mod_autoindex.so
#LoadModule buffer_module modules/mod_buffer.so
...
```

We will need to make sure that we're adding a few modules for secure proxy authentication. In `mod_proxy`, AJP, instead of the older AJK module, that takes care of the AJP connection to Tomcat and `mod_ssl`, you'll find these modules listed somewhere in this list—just make sure that you uncomment them, assuming that they've already been built for the package that you're using.

The `httpd` file includes some conditional directives for loading modules. So if the `ssl_module` has been included, then the directives are run as shown in the following code snippet:

```
<IfModule ssl_module>
  #Include conf/extra/httpd-ssl.conf
  Include conf/extra/httpd-ahssl.conf
  SSLRandomSeed startup builtin
  SSLRandomSeed connect builtin
</IfModule>
<IfModule http2_module>
  ProtocolsHonorOrder On
  Protocols h2 h2c http/1.1
</IfModule>
```

One of them is `Include conf/extra/httpd-ahssl.conf`, which we will be configuring in the next section.

We'll use Tomcat as the servlet container rather than the default embedded jetty. The following is a well-documented configuration for using Tomcat with Apache HTTPD:

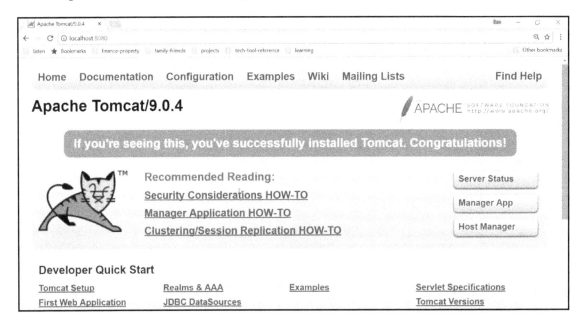

It is necessary to set the `Catalina_home` environment variable to the base directory of Tomcat. If you don't see an obvious error message on the Tomcat startup, you'll know that this environment variable needs to be set.

To install GeoServer on Tomcat, download a `.war` file from the GeoServer site and move it into the `web apps` directory. It should be expanded into the familiar directory structure when Tomcat run with the WAR for the first time. If this fails, you may need to check your privileges. The main configuration for Tomcat is the `server.xml` file under Tomcat's `conf` directory. The main thing we need to pay attention to is the AJP connector. Here, you can see that the connector is listening at `port 8009`, as shown in the following code:

```
<!-- Define an AJP 1.3 Connector on port 8009 -->
<Connector port="8009" protocol="AJP/1.3" redirectPort="8443" />
```

This will be important as we complete proxy configuration in the next section. Before we do that, let's take a look at the hosts file. In Windows, this is available under the `C:\Windows\System32\drivers\etc` path. In Linux, it is usually under the `etc` directory. This file allows you to map a hostname to an IP address. You can see here that the familiar `localhost` is mapped to our local IP address, and that's how it can work on the OS, as shown in the following snippet:

```
#localhost name resolution is handled within DNS itself.
   127.0.0.1 hello.world
   127.0.0.1 localhost
#::1 localhost
```

I've added the hostname `hello.world`, and I'll be using this throughout the section, but you may want to set a hostname here that is equivalent to your web host if you want to do local testing.

In this section, you learned about the proxy environment and configuration of hosts to facilitate TLS. In the next section, we'll move on to HTTPS, and you'll learn how to finish the configuration of TLS and secure proxy.

HTTPS with TLS and certificates

In the previous section, you learned about the underlying software for the proxy platforms HTTPD and Tomcat. There are `config` files and the `OS host` file. In this section, you'll learn the steps to complete the secure proxy platform. HTTPS is the standard method to securely send and receive messages on the web. You'll learn about TLS and certificates, and how to create a self-signed certificate for testing. You'll configure the certificate to be leveraged by the server for HTTPS. Finally, you'll configure client software to trust the certificate and test secure communication.

HTTPS is very widely used on the internet, since any text sent over HTTP will be transmitted in plain text and can be intercepted. The Wikipedia page uses HTTPS, because when we use it, we send searches, and the searches could be intercepted if they're sent in plain text.

HTTPS uses a protocol called TLS, which superseded SSL. The following `openssl` command provides information about a certificate:

```
openssl x509 -text -noout -in C:\Apache24\conf\ssl\geoserver.crt
```

TLS is often described as a handshake between a client and server. Certificates granted by owners are signed by a certificate authority with a private key, an encoded string known only to the authority and identified by a readable public key that is included with the certificate. Certificates contain information about the owner—most importantly, the hostname that the certificate applies to. This is listed under **CN** or common name; you can see the **CN** in the following screenshot:

```
C:\WINDOWS\system32>openssl x509 -text -noout -in c:\Apache24\conf\ssl\geoserver.crt
Certificate:
    Data:
        Version: 3 (0x2)
        Serial Number:
            b4:ba:54:6d:ea:1f:f2:44
    Signature Algorithm: sha256WithRSAEncryption
        Issuer: C=AU, ST=Some-State, O=hello world, OU=hello world, CN=hello.world
        Validity
            Not Before: Feb  6 22:10:34 2018 GMT
            Not After : Feb  6 22:10:34 2019 GMT
        Subject: C=AU, ST=Some-State, O=hello world, OU=hello world, CN=hello.world
        Subject Public Key Info:
            Public Key Algorithm: rsaEncryption
                Public-Key: (2048 bit)
                Modulus:
                    00:bd:e1:71:91:c8:64:06:7f:c4:fe:f0:4f:83:0e:
                    c6:69:92:70:fa:29:e4:50:8a:8c:09:f6:46:db:f0:
                    c6:97:31:f2:f5:fc:8d:24:38:76:54:47:54:84:94:
                    70:90:4c:3e:e7:d3:64:a7:a9:5d:0e:5c:1e:11:3e:
                    09:71:9a:e7:3c:45:90:60:f6:ca:59:87:4b:e5:ec:
                    12:f2:c6:1f:1b:e9:ce:c0:9e:d6:f3:df:43:c2:84:
                    74:f3:81:d3:28:5c:52:2e:f3:84:25:9d:08:55:43:
                    33:29:cd:7a:bd:f9:d5:0a:af:77:58:2b:16:53:ea:
                    f4:27:97:cc:8f:3c:88:9b:93:c7:98:13:dc:18:5b:
                    ea:10:c1:5c:1b:05:f0:ff:aa:f5:fe:46:ce:12:73:
                    b4:ad:de:7d:11:22:8d:17:c0:97:d8:9e:4d:fb:74:
                    60:94:92:e5:b5:77:77:b4:b6:3f:a1:eb:83:fe:37:
```

A certificate must be accepted by a client to complete the handshake. Clients will automatically accept certificates signed by authorities that are included on the list of trusted authorities for that client. Root authorities, since they are on the original signer's certificates. Clients are skeptical of self-signed certificates that are not known from among the well-known root authorities. We will be self-signing a certificate to test HTTPS; therefore, we are acting as our own root authority. Client software will be skeptical and may produce warnings and errors.

The following is the `openssl` command to create a new certificate and a private key:

```
openssl req -x509 -nodes -days 365 -newkey rsa:2048 -keyout
C:\Apache24\conf\ssl\geoserver-demo.key -out
C:\Apache24\conf\ssl\geoserver-demo.crt
```

Despite the name, the open SSL executable is still widely used to generate certificates for TLS. Open SSL is often distributed with Apache HTTPD, as is the case with the `Apache house` package I am using. With `openssl` on the OS path, or `run` from the correct directory, the preceding command will generate a new private key and self-signed certificate with certain options. You will be prompted for `x509` information, which is the standard set of information about the server and the owner. We just saw some of this information a moment ago. Again, the only field that's really important to us is the common name, which is the server hostname, and that can be set to whatever hostname you used in the `hosts` file, such as `hello.world` in our example. This particular certificate will be valid for `365` days. A private key file that's nonpassword-protected will be created for the server to use. This will be generated with RSA encryption to a length of `2048` bits. When you are doing this with an actual certificate authority, you will just create the requests, so you'll have a different set of options here, and you will then send that request to the certificate authority to sign. In our case, the private key should only be readable to root and admin, and the certificate should be readable by everyone, since the certificate will be used to validate the certificate that we'll be signing. So, if you run the preceding command, you'll start to see prompts for all of those `x509` data fields, as shown in the following screenshot:

```
C:\Users\mearns>openssl req -x509 -nodes -days 365 -newkey rsa:2048 -keyout c:\Apache24\conf\ssl\geoserver-demo.key -o
ut c:\Apache24\conf\ssl\geoserver-demo.crt
Loading 'screen' into random state - done
Generating a 2048 bit RSA private key
......................+++
.................................+++
writing new private key to 'c:\Apache24\conf\ssl\geoserver-demo.key'
-----
You are about to be asked to enter information that will be incorporated
into your certificate request.
What you are about to enter is what is called a Distinguished Name or a DN.
There are quite a few fields but you can leave some blank
For some fields there will be a default value,
If you enter '.', the field will be left blank.
-----
Country Name (2 letter code) [AU]:
State or Province Name (full name) [Some-State]:
Locality Name (eg, city) []:
```

As you saw in the previous section, a separate `conf` file is included in the `main conf` file and is used to configure TLS, akka SSL. There are two options of `SSL conf` files that we can use. We are using the default `httpd-ahssl.conf` file. This file is configured for SNI, which allows multiple certificates for multiple hostnames to be used at a single physical host IP address. We are not utilizing SNI here, but that's something you might want to consider. Sections for additional hosts will be inserted in here; you can see them commented in the previous code screenshots, and in the following snippet:

```
<VirtualHost *:443>
  SSLEngine on
  ServerName hello.world:443
  SSLCertificateFile "${SRVROOT}/conf/ssl/geoserver.crt"
  SSLCertificateKeyFile "${SRVROOT}/conf/ssl/geoserver.key"
  DocumentRoot "${SRVROOT}/htdocs"
  CustomLog "${SRVROOT}/logs/ssl_request.log" \
 "%t %h %{SSL_PROTOCOL}x %{SSL_CIPHER}x \"%r\" %b"
  <Directory "${SRVROOT}/htdocs">
    Options Indexes Includes FollowSymLinks
    AllowOverride AuthConfig Limit FileInfo
    Require all granted
  </Directory>
  ProxyPreserveHost On
  ProxyRequests Off
  ProxyPass /geoserver ajp://hello.world:8009/geoserver
  ProxyPassReverse /geoserver https://hello.world/geoserver
</virtualhost>
```

The certificate key and server name have all been edited to reflect current paths and the server name, as shown in the preceding snippet. Note `port 443`, which is the default SSL port. You must also add the directives the `ProxyPass /geoserver ajp://hello.world:8009/geoserver` and `ProxyPassReverse /geoserver` regarding the proxy. `Port 8009` refers to the Tomcat AJP connector port shown in the Tomcat server XML configuration that you saw in the previous section.

We will run the following `curl` command to test our certificate against the server:

```
curl --ssl -v --cacert C:\Apache24\conf\ssl\geoserver-demo.crt
https://hello.world | less
```

Of course, you'll need to restart Apache and Tomcat to enable the configuration that you just made on the last file. We're using the `less` command for output, but you can see in the following screenshot that the `TLS handshake` is actually being logged between the client and server, the exchange of certificates, and so on:

```
Command Prompt
ort 443 (#0)
* ALPN, offering h2
* ALPN, offering http/1.1
* successfully set certificate verify locations:
*    CAfile: C:\Apache24\conf\ssl\geoserver.crt
  CApath: none
} [5 bytes data]
* TLSv1.2 (OUT), TLS handshake, Client hello (1):
} [209 bytes data]
* TLSv1.2 (IN), TLS handshake, Server hello (2):
{ [108 bytes data]
* TLSv1.2 (IN), TLS handshake, Certificate (11):
{ [937 bytes data]
* TLSv1.2 (IN), TLS handshake, Server key exchange (12):
{ [333 bytes data]
* TLSv1.2 (IN), TLS handshake, Server finished (14):
{ [4 bytes data]
* TLSv1.2 (OUT), TLS handshake, Client key exchange (16):
} [70 bytes data]
* TLSv1.2 (OUT), TLS change cipher, Client hello (1):
} [1 bytes data]
* TLSv1.2 (OUT), TLS handshake, Finished (20):
} [16 bytes data]
* TLSv1.2 (IN), TLS handshake, Finished (20):
{ [16 bytes data]
* SSL connection using TLSv1.2 / ECDHE-RSA-AES256-GCM-SHA384
* ALPN, server accepted to use http/1.1
* Server certificate:
*   subject: C=AU; ST=Some-State; O=hello world; OU=hello world; CN=hello.world
```

The less command is used for paging information, and we're expecting to get a lot back, so that's why this less command is here after the pipe in the following screenshot:

```
C:\Users\mearns>curl --ssl -v --cacert C:\Apache24\conf\ssl\geoserver.crt https://hello.world | les
s
* Rebuilt URL to: https://hello.world/
  % Total    % Received % Xferd  Average Speed   Time    Time     Time  Current
                                 Dload  Upload   Total   Spent    Left  Speed
  0     0    0     0    0     0      0      0 --:--:-- --:--:-- --:--:--     0*   Trying 127.0.0.1.
..
* TCP_NODELAY set
* Connected to hello.world (127.0.
C:\Users\mearns>
```

The more command is what you would use in Linux.

To test this out on the browser, we can go to the `hello.world` hostname that we checked out in the last section, but we'll add HTTPS. If you haven't already added the certificate to trusted root authorities for this particular client or for your OS, you'll get a warning error message like the one shown in the following screenshot:

You can skip through, but we will instead add a certificate for the server—which in our case is the computer that we're using—to our list of trusted root authorities. We can do so in Chrome by clicking on **Certificate** to see more information about the certificate:

You can copy the certificate to a file, as shown in the following screenshot, saving it as **PKCS #7**, which is just an encryption standard:

You can then save it to a file. Then, in Chrome's **Settings** section, under the **Advanced** section, you'll find **Manage certificates**, as shown in the following screenshot:

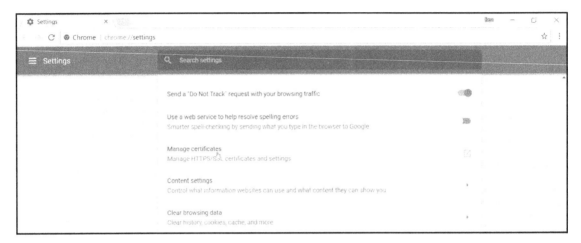

Later, you can import the certificate **PKCS #7**. Here's the certificate we created previously in the browser:

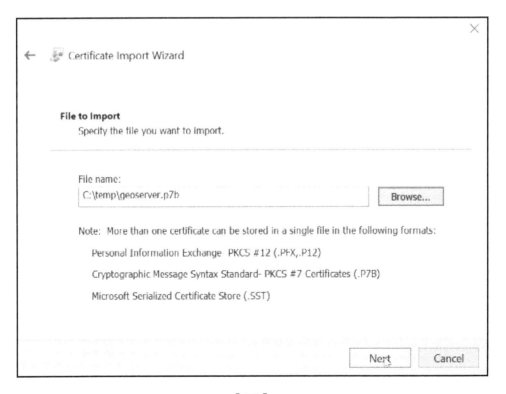

We will now add this not to the **Personal** certificates, but to the **Trusted Root Certification Authorities**:

After clicking on **OK**, and after a few more clicks, we will have successfully imported the certificate. Our self-signed certificate will now be using a trusted authority, so we could then enter the HTTP endpoint for our hosts, and we could verify that the TLS is working by checking out the **Developer tools** and, under the **Security** tab, we will see that we do have a secure connection using TLS, as shown in the following screenshot:

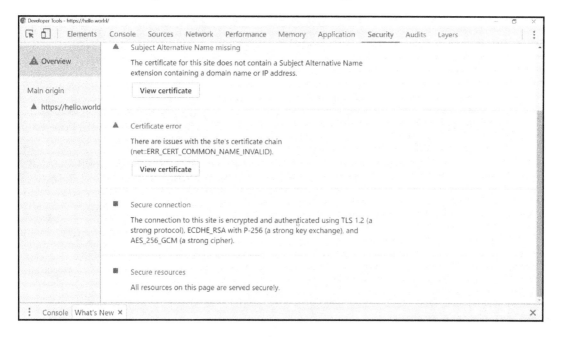

In this section, you learned how to configure certificates on the server and clients, and test the secure communication with HTTPS. In the next section, you'll learn about GeoServer concepts for permission management and authentication, and configure GeoServer to protect sensitive data services.

GeoServer authentication

In the previous section, you learned how to configure HTTPS and certificates for secure communication between client and server over TLS. In this section, you'll learn how to configure GeoServer's permission settings for authentication. First, you'll learn how to set up users, roles, and groups for GeoServer. Next, you'll learn about authentication providers to configure the way that users are authenticated to GeoServer. Finally, you'll learn about data and using service-level security to limit access to sensitive actions and resources.

In the GeoServer web administration interface, there's a variety of settings that can be accessed under the **Security** section, as shown in the sidebar in the following screenshot:

In the **Users, Groups, Roles** section, you can ignore the **Services** tab for our purposes. Now, going into **Users/Groups**, Users obviously refer to the logins that different individuals would be using, or that applications would use. I've just set up **admin** as the default user of GeoServer, as shown in the following screenshot:

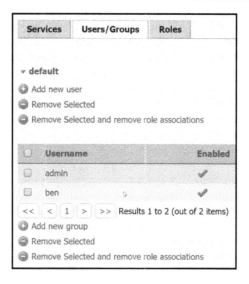

The **ben** user is one that I just created, and of course you'll be creating a password along with the login. You can also set groups and roles here:

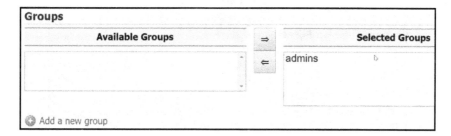

You may also need to create a group; I've created a group called `admins`, as shown in the preceding screenshot, and this `admins` group includes the roles of `ADMIN` and `GROUP_ADMIN`, as shown in the following screenshot:

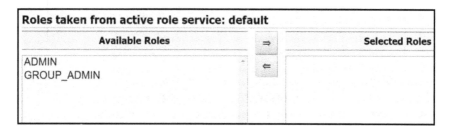

They're not set here because they're set through the group. So you could set these roles directly to the user, or you can create a group and then select the roles for that group. These roles are related to sets of permissions that are not directly set here. The following is just kind of a placeholder:

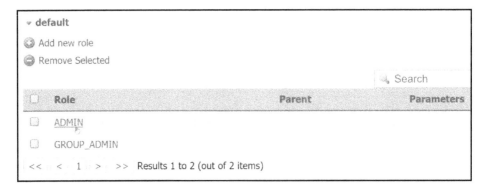

The permissions for different groups will be set under the **Service and Layer Level Security Settings**, which we'll cover towards the end of this section. In the **Authentication** section, you can see the following **Authentication Filters** section:

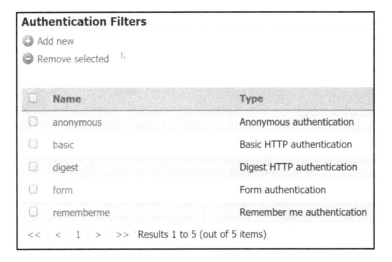

This section deals with the methods of authentication that are permitted and used for GeoServer. These are the default filters, except for **digest**, which we added. I hadn't had much luck with **digest** authentication; I think there may be a bug in this version, but hopefully that will be fixed in future patches of GeoServer.

So instead, we'll be using basic authentication. Form authentication is another filter that you may want to use, and there are other filters that you can also enable, such as filtering by using Apache sessions, which would probably be more secure than what we're doing, but also more complex and out of the scope of this book. The filter chains relate to the different sections of URLs—so when you get, for example, `localhost:8080/geoserver` or `hello.world/geoserver/rest`, for anything, then for anything after that you'll be directed to the rules for this REST to chain.

We're going to be using WFS in this section, so I've set up a WFS filter chain:

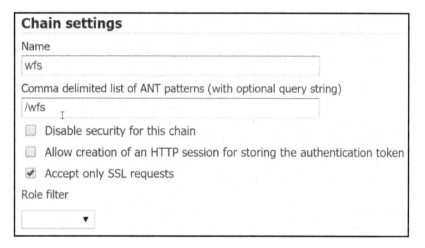

Any WFS query will be captured by this, and the authentication methods that are allowed will move over to the **Selected** area by default, as shown in the following screenshot:

We'll capture any requests that are not captured by other filter chains, and **digest**, **basic**, and **anonymous** are all allowed here. So for whatever reason, if **basic** fails because the login and password aren't correct, then the user will be allowed to access these resources anonymously. This is, of course, something that you won't want to allow if you have sensitive resources that you don't want an **anonymous** user to gain access to.

You should also note that there are HTTP methods that are used to filter, so a POST to WFS will be acted on differently than a GET request to HTTPS or to WFS. Additionally, the **Provider Chain** section shown in the following screenshot relates to the data that stores the actual username and password; you can just kind of ignore that for our purposes:

Finally, we can set data or layer access permissions under the **Data Security** section, as shown in the following screenshot:

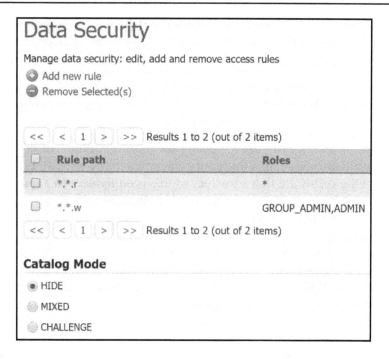

The service-level permissions would be under the **Services** section. Our bases should be covered with that WFS filter, but just as a matter of precaution, for any action that makes a hard right to GeoServer, the GROUP_ADMIN and ADMIN roles are required. You can see the roles in action in the preceding screenshot. These are just placeholders from the **User** section, but users are tied to these roles, so our new **ben** user, for example, would be allowed to make a request that writes. In this section, you learned about GeoServer authentication identity and permissions management. In the next section, you'll learn how to apply GeoServer access management to an OpenLayers web app for secure authentication.

Secure login from OL

In the previous section, you learned about the GeoServer settings for authentication, such as authentication based on users, groups, roles, filters, and chains. In this section, you'll learn how to securely log in to a configured GeoServer instance with an OpenLayers-based web app. First, you'll learn how to test authentication with the command-line utility curl. Next, we'll make some minor modifications to some OL code that was previously used to make non-HTTPS GeoServer WFST requests. Finally, we'll test the concepts you learned about in this section.

Make sure Apache HTTPD and Tomcat are started, if they haven't already been started. If GeoServer is not available at the Tomcat address `localhost:8080/geoserver`, then check permissions and run again. Now, we will test our secure login with curl:

```
curl -v --cacert C:\Apache24\conf\ssl\geoserver.crt -u admin:geoserver -v -
H "Content-type: text/xml" -d @update.xml https://hello.world/geoserver/wfs
```

One of the options you can see in the preceding command is `-v`, to give us, for both outputs, `--cacert`, which we can use to specify the location of the trusted server certificate. This could also be set to a folder of certificates that you trust, but here we can just directly select the certificate that's trusted, so this is equivalent to a root authority certificate and a user. We're going to use basic authentication with the `admin` GeoServer user. This could also be the user that we created in the last section. Then, the header is going to include the content type `text/xml` to let the server know what content is being sent. The content is an XML file, `@update.xml`—which is equivalent to something that you might get from the demo requests that are included with GeoServer for doing an update via WFST—and finally the HTTPS URL of the WFS service. Running this, we can see first of all that our verification of the location of the certificate was successful, as shown in the following screenshot:

```
C:\Users\mearns\Google Drive\projects\packt\LearningGeoserver\volume3\Section 4>curl -v --cacert C:
\Apache24\conf\ssl\geoserver.crt -u admin:geoserver -v -H "Content-type: text/xml" -d @update.xml
https://hello.world/geoserver/wfs
*   Trying 127.0.0.1...
* TCP_NODELAY set
* Connected to hello.world (127.0.0.1) port 443 (#0)
* ALPN, offering h2
* ALPN, offering http/1.1
* successfully set certificate verify locations:
*   CAfile: C:\Apache24\conf\ssl\geoserver.crt
  CApath: none
* TLSv1.2 (OUT), TLS handshake, Client hello (1):
* TLSv1.2 (IN), TLS handshake, Server hello (2):
* TLSv1.2 (IN), TLS handshake, Certificate (11):
* TLSv1.2 (IN), TLS handshake, Server key exchange (12):
* TLSv1.2 (IN), TLS handshake, Server finished (14):
* TLSv1.2 (OUT), TLS handshake, Client key exchange (16):
* TLSv1.2 (OUT), TLS change cipher, Client hello (1):
* TLSv1.2 (OUT), TLS handshake, Finished (20):
* TLSv1.2 (IN), TLS handshake, Finished (20):
* SSL connection using TLSv1.2 / ECDHE-RSA-AES256-GCM-SHA384
* ALPN, server accepted to use http/1.1
* Server certificate:
*   subject: C=AU; ST=Some-State; O=hello world; OU=hello world; CN=hello.world
*   start date: Feb  6 22:10:34 2018 GMT
*   expire date: Feb  6 22:10:34 2019 GMT
*   common name: hello.world (matched)
*   issuer: C=AU; ST=Some-State; O=hello world; OU=hello world; CN=hello.world
```

The `TLS handshake` runs through successfully. Then finally, we get a response from the server that lets us know that the update was successful, as shown in the following screenshot:

```
<?xml version="1.0" encoding="UTF-8"?><wfs:WFS_TransactionResponse version="1.0.0" xmlns:wfs="http:
//www.opengis.net/wfs" xmlns:ogc="http://www.opengis.net/ogc" xmlns:xsi="http://www.w3.org/2001/XML
Schema-instance" xsi:schemaLocation="http://www.opengis.net/wfs https://hello.world:443/geoserver/
chemas/wfs/1.0.0/WFS-transaction.xsd"><wfs:InsertResult><ogc:FeatureId fid="none"/></wfs:InsertRes
lt> <wfs:TransactionResult> <wfs:Status> <wfs:SUCCESS/> </wfs:Status> </wfs:TransactionResult></wf
:WFS_TransactionResponse>* Connection #0 to host hello.world left intact
```

Now, we will do the same thing in OpenLayers. First, we need to make a few modifications to the OpenLayers code that may have been used previously to do WFST with a non-HTTPS URL endpoint; anywhere where this HTTP appears could be a place where it's trying to contact the server with non-HTTPS. You can see in the following code that we've modified this to show the correct HTTPS endpoint:

```
url = 'https://hello.world/geoserver/wfs';
```

We also made another modification here as well:

```
var host = 'https://hello.world/geoserver';
```

 Use *Ctrl* + *f* in the `wfst-ol html` file to find the preceding code, which is provided along with the code files of this book.

The XML that's going to be generated here is equivalent to what we just sent in the curl request. It is just generated programmatically using data from the OpenLayers app, so let's finally check out the OpenLayers app.

The following screenshot shows the OpenLayers app, and, when clicking on one of these features, we are prompted to authenticate:

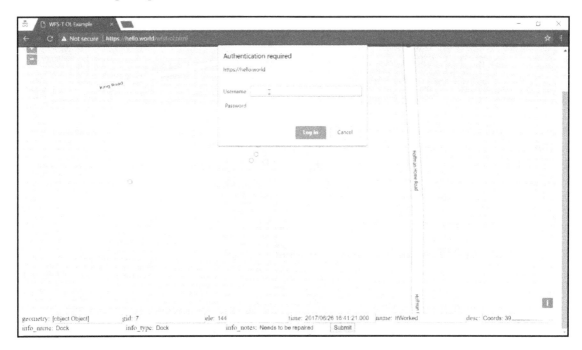

So, in our GeoServer authentication settings, everything is working correctly, and we get a successful response back, as shown in the following screenshot:

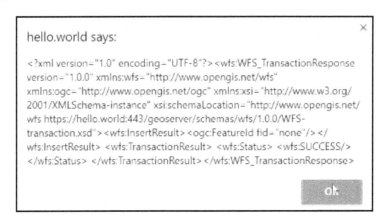

The response can be handled in a different way, of course; this is just an example. But if we refresh the browser, we can see that the change has been persisted and saved to the database. The change is shown in the following screenshot:

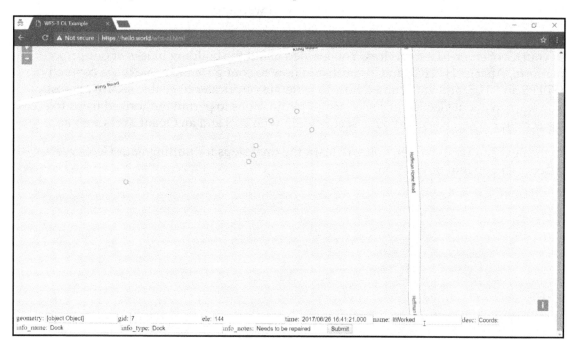

This is the final section of this chapter, in which you've learned how to secure GeoServer using authentication. You learned how to set up a proxy platform with Apache HTTPD and Tomcat. You created and used certificates to establish a trusted HTTPS connection between the client and server. You configured GeoServer with user-access permissions and authentication filter chains. You've added HTTP services to an OpenLayers app for secure, authenticated write access.

Summary

Great job on working through this difficult section! You can take the fundamentals you learned here to create a really great authenticated web app.

In this chapter, you learned to use HTTPS to provide secure login from an OpenLayers app to your GeoServer-based services. You learned about the building blocks of our proxy platform, Apache HTTPD, and Tomcat, and how to configure them. Next, we covered HTTPS and TLS, and you learned how to generate certificates to enable secure requests. You learned about GeoServer's user-access permissions to permit authorized users to access the server. Finally, we tested out a secure login with curl and an OpenLayers web app.

In the next and final chapter, you will learn the final steps for putting your GeoServer-based web app into production.

Putting it into Production

5

In this final chapter, you will learn how to get your GeoServer instance running on the web for your end users, and how to keep it running. First, you'll learn about the hosting options for GeoServer and how to deploy. You'll learn how to monitor GeoServer to maintain a stable instance and troubleshoot. Next, you'll learn about backup and recovery techniques, and finally, we'll go through a checklist of settings for production of GeoServer to make sure you have your bases covered.

In this chapter, we will cover the following topics:

- Hosting your GeoServer instance and app
- Monitoring the GeoServer instance
- Backup and recovery
- Production checklist

Hosting your GeoServer instance and app

In this section, you'll learn about the three main options for GeoServer hosts; hosting on-premises, cloud-based virtual machines, and containerized virtualization. For each of these options, you'll learn the advantages and disadvantages and the basic steps involved in deployment. In some ways, the most straightforward method for hosting is with the server at your physical location: on-premises hosting. It is possible to host directly on the same machine you used to do development using a server that is dedicated to hosting web resources, or to use a virtual machine that is created on a local data center; for instance, if your organization has a VMware data center cluster. With a virtual machine, you can start small, do benchmarking, and scale up if needed. A physical server does not afford such flexibility, so you should exceed the expected load when provisioning resources. An on-premises server may be the most secure of all options if firewalls and other security precautions are taken.

If the main load on the server will be from local machines, you may also get the benefit of faster performance, and there may also be greater tolerance for server downtime if users tend to be within your organization and you're doing maintenance outside of usual business hours.

A typical configuration will have the OpenLayers web application or another mapping app running on Apache HTTPD at port 80, with or without proxying to Tomcat via HTTPS, or just directly referencing Tomcat with HTTP at port 8080:

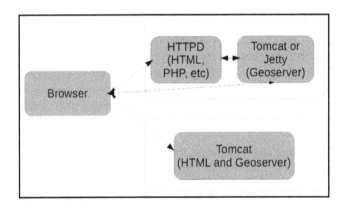

It is also possible to host the web app solely on Tomcat, and this may be a good option if no third-party server-side scripting, such as PHP, is required. It is harder to host HTML pages with Jetty, so it's best to use Tomcat or Tomcat with HTTPD if you are hosting the web app and GeoServer on a single mode. If you are serving external users on the web, you may need to configure the organizations, firewall to allow connections to the ports that GeoServer uses. The port for Postgres does not need to be open to the outside – it should not be open to the outside, as GeoServer can communicate with Postgres locally or on a local network address permitted through the firewall and Postgres config files. A dedicated public address may also need to be provisioned, and/or a public domain registered with the server may need to be registered to that domain or a subdomain. Having a local instance of GeoServer for staging and another for production is a good idea regardless of the production server location, on or off premises. You will learn how changes from staging to production can be propagated with the backup and restore plugin in the *Backup and recovery* section of this chapter.

A virtual machine directly analogous to the server, or an on-premises virtual machine, can also be created on cloud infrastructure-as-a-service platforms such as Amazon EC2, Microsoft Azure, Google Cloud Platform, and so on. The following screenshot shows the web administration console:

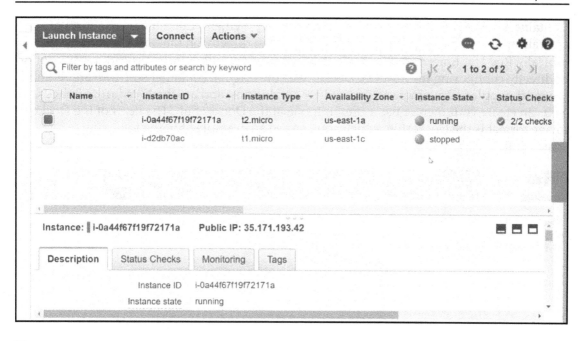

Here, you can create new instances, manage them, and reconfigure the size and resources available to each. All the major providers also provide a command-line interface executable for completing these tasks or scripting them. A major benefit of virtual machines is the ability to scale up or scale down resources on demand. Want an extra-fast CPU? Just add it through the console. More memory? Same thing. Once the virtual machine has been started, the process for installing and configuring GeoServer, Postgres, and other server components is almost identical to doing so on your local machine, except that you will need to connect and manage the server through SSH remote desktop, or something similar. In fact, physical-to-virtual tools allow you to spin up a virtual machine identical to your physical machine in the cloud. You will probably also need to configure firewall and identity and permissions in the console. Another big benefit with cloud-hosted IS is the relative ease of creating clusters. Not only can multiple identical VMs be spun up in a fraction of the time of physical modes, but the major IS performer providers all have middleware accessible via the console to help create clusters for high availability and performance.

Containers are yet another option for GeoServer deployment. Containers are a form of virtualization, yet, unlike virtual machines, they run directly on the host OS kernel – typically Red Hat Enterprise Linux. Containers provide neither the overhead nor the option of directly configuring the host OS. Docker was the original container software and is still the most popular. Containers are created for single apps, and are deployed from images. You can create your own or use a pre-built one at the Docker hub. The heart of the Docker image format is the Dockerfile, which you can see an example of in the following screenshot:

The Docker image and Dockerfile are for a GeoServer Docker container, which looks just like a shell script running all the commands necessary to create the container. When Docker creates a container for the first time, Docker caches files to build all subsequent containers very quickly. Containers have even more flexibility for on-demand scaling and clustering for the best end-user experience. In this section, you learned the major deployment options available for GeoServer. There are other options too. Software-as-a-Service providers such as AQGO offer a managed GeoServer instance where you don't need to do anything to install it, or do much to manage it. This can work for any use case. In the next section, you'll learn how to monitor your GeoServer instance to make sure it is delivering the best performance.

Monitoring the GeoServer instance

In the previous section, you learned about hosting options for your GeoServer instance. In this section, you'll learn how to make sure the instance remains healthy, and how to detect and remedy errors if something goes wrong. You'll learn about logging and all the different logs associated with GeoServer and other app components. You'll learn how to use the GeoServer monitoring plugin. Finally, we'll cover some other software for monitoring the infrastructure supporting our apps.

Server software creates log files to provide administrative insight into the operation of the software. The server may use a single log file, or it may use additional log files to divide up categories of events – access versus errors, for instance. While the access log could be used to audit server traffic, the error log could be used to diagnose and troubleshoot problems. Different log levels can also be configured after the stage of the server development. Sparse logs do not hinder server performance in a production environment, while verbose logs are helpful for testing in a development environment. In addition to the single log file, .log, stored in the logs directory under the GeoServer data directory, numbered log files are automatically created by log rotation, so a big log file does not become a drain on GeoServer resources.

Logs are configured in the web administration interface or via plain text configuration files. logging.xml is located directly under the data directory. This file corresponds to the logging section of the global settings page, which you can see in the web administration interface, shown in the following screenshot:

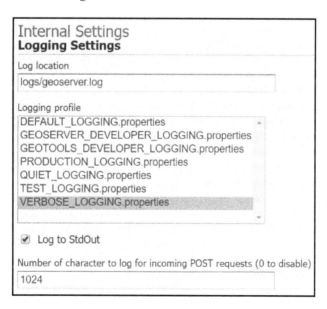

You can select debug levels and log profiles here. Here is one of the profile configuration files, which controls log level and format, as shown in the following code snippet:

```
log4j.rootLogger=WARN, geoserverlogfile, stdout

log4j.appender.stdout=org.apache.log4j.ConsoleAppender
log4j.appender.stdout.layout=org.apache.log4j.PatternLayout
log4j.appender.stdout.layout.ConversionPattern=%d{dd MMM HH:mm:ss} %p
[%c{2}] - %m%n

log4j.category.log4j=FATAL
log4j.appender.geoserverlogfile=org.apache.log4j.RollingFileAppender
# Keep three backup files.
log4j.appender.geoserverlogfile.MaxBackupIndex=3
# Pattern to output: date priority [category] - message
log4j.appender.geoserverlogfile.layout=org.apache.log4j.PatternLayout
log4j.appender.geoserverlogfile.layout.ConversionPattern=%d %p [%c{2}] -
%m%n

log4j.category.org.geotools=TRACE
log4j.category.org.geotools.factory=TRACE
log4j.category.org.geotools.renderer=DEBUG
log4j.category.org.geoserver=TRACE
log4j.category.org.vfny.geoserver=TRACE
```

These are under the `logs` directory. A single profile has been selected in `logging.XML`. With profiles, you can really drill down to the information that is important to you and present it in a format that matches your workflow needs. Note that log levels are set for the underlying components in the GeoServer software. This is very helpful for debugging. The number of rotated log files to be retained is also set, so you can see a number of log files retained, as shown in the preceding screenshot, and each of these components is a GeoServer component. Finally, the `web.xml` file under the `WEB-INF` directory includes a parameter, log request bodies, that can be set to true, along with the logging filter, in order to enable logging of requests. Logging is useful for understanding what potentially important information related to web app and server behavior is also logged for Tomcat, and Apache HTTPD, and any additional software components such as PHP.

While logging is useful for understanding what server software components are doing, the monitor extension is a great tool for following HTTP requests through their life cycle on GeoServer, as shown in the following screenshot:

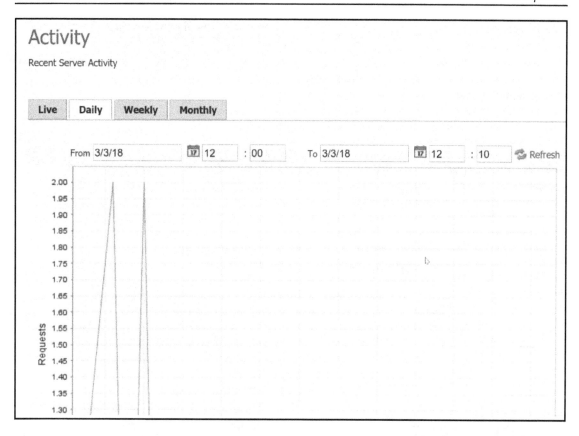

The monitor extension is more dynamic than logging. It provides an API that provides live monitoring subsets, for instance, data graphs, and can also write to the database.

By default, only certain requests are monitored. Many filter parameters are available; for example, only WFS requests within a certain extent and from a certain time period, so that you can hone in on the kinds of requests you're interested in. Finally, one of two modes is chosen; first, there is live mode, which stores a number of recent requests in memory. This is the fastest option, and is used for monitoring current performance. There is also history mode, which is slower but crucial for monitoring the performance of many requests over time.

In addition to monitoring and logging software behavior, it is also a good idea to monitor the infrastructure on which the software is running. This may be important, for example, to respond to a server that has crashed or is coming close to hitting a bottleneck in a particular kind of resource. The major IS providers offer monitoring software, such as Amazon CloudWatch, which you can see here:

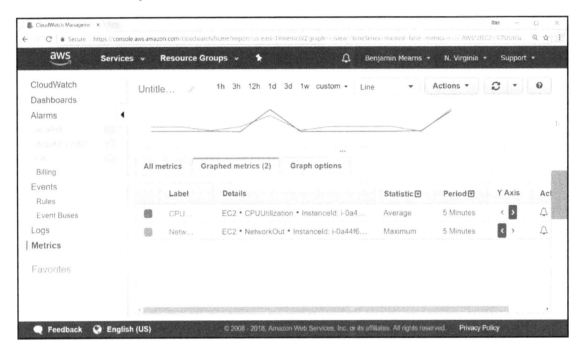

CloudWatch is relatively easy to set up for your Amazon VMs or cluster, and can trigger custom behaviors or send custom alerts if the server is exhibiting notable behavior.

Nagios is an industry standard in infrastructure monitoring. It is a mature open-source project with many extensions and capabilities. Nagios is largely equivalent to CloudWatch, though configuration is more difficult, and it is appropriate for public and private clouds and not tied to a particular IS provider. In this section, you learned how to find the current and historical status of your GeoServer software stack and infrastructure. In the next section, you'll learn how to set up a backup for GeoServer and how to recover when it's needed.

Backup and recovery

In the previous section, you learned how to stay informed about your current and historical server status. In this section, you'll learn how to use the backup and restore plugin along with reoccurring scheduled tasks to keep your GeoServer instance stable, current, and backed up. First, we'll look at the GeoServer backup and restore plugin. You'll learn how to schedule backups and how to restore them. Next, you'll learn about a unexpected use of backup and restore, for pushing staging server changes to production, cluster nodes, and standby nodes. Finally, you'll learn about the scheduled tasks that will keep your server backed up, in sync, and stable.

The backup and restore plugin provides a managed way to restore changes to a GeoServer instance, along with messages to help troubleshoot if anything goes wrong. After the plugin is installed, you will have access to the following page in the web administration interface:

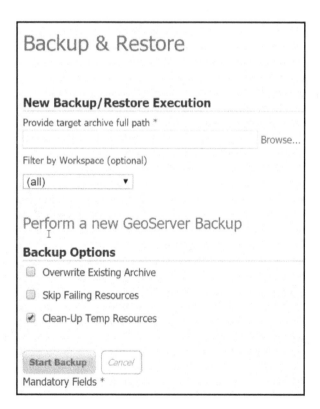

To use the plugin, just specify a location on which to write the file – ideally an external disk that will not fail if the active GeoServer node does. Recovery is also straightforward. Locate the backup file and start, the restore by clicking on the **Start Restore** button:

Dry-Run is useful for testing to make sure failures do not occur, as this will not alter the GeoServer instance. If failures occur with certain resources, for example, a data file that is no longer found, failing resources can be skipped. The plugin only stores configuration file data, so all non-configuration data must be backed up separately.

The backup and restore plugin can also be used to synchronize changes between staging and production, nodes in a cluster, or a standby node:

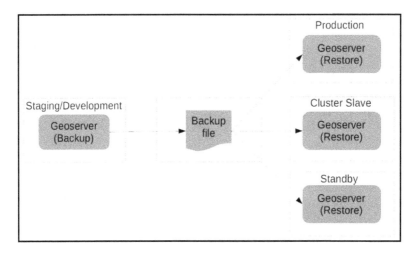

A staging node would be a local GeoServer instance, for example, one found on a developer's laptop, on which changes are made. Other nodes would have a read-only configuration, and configuration changes would only propagate from the staging server. To do this, just run backup on the staging machine and then use the backup file to run restore on the nodes to be updated.

Backing up is just one of the tasks that can be scripted and run automatically. Software to run automated tasks includes **Task Scheduler**, which is on Windows—you can see the interface in the following screenshot—and cron, which is on Linux:

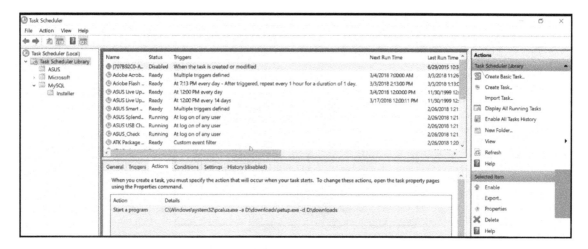

Tasks to run automatically on a schedule for GeoServer include backup, with the backup and restore plugin API and curl. You'll note that these scheduled tasks can run anything that you can put in the command line, or you can even just create a batch file or shell script from the command line, and then just run that with the scheduled task software. And curl, being a command, can be run in any of these scripts, and is useful if you're interacting with any kind of REST API or HTTP. GWC caching via curl can also be done since it has a REST API. You can copy logs out of a `data` directory and back them up to a different external directory. You can also include the naming, and in the naming you can include the current date and time.

Restarting GeoServer regularly is a good idea. Backing up the `data` directory and pushing these changes to other nodes can be done with rsync, which is a great piece of software on Linux. This backs up changes to files and chunks and is very fast, and good rsync alternatives for Windows include syncthing and CW rsync, which is rsync in a cygwin wrapper.

Database backup dumps should also be automated; database backup can be done manually in pgAdmin using the backup command, as shown in the following screenshot:

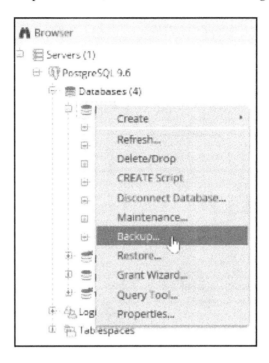

If you use that command, you'll see the command-line options that are used along with the command, and it's actually running pg_dump, which is the command. You can see the pg_dump along with all of the options that are used on that command in the following screenshot:

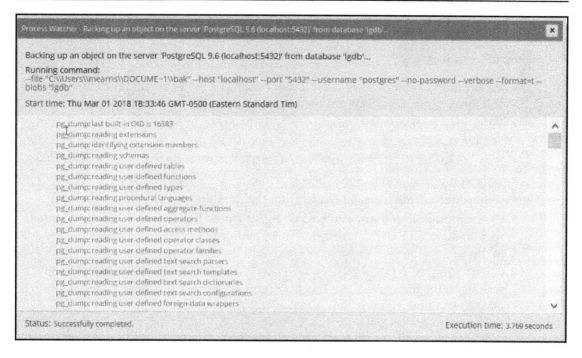

So, you can always set up your backup first on pgAdmin, with the options that you wish to use through their GUI, and then you can get the command-line options below the **Running command:**, and that makes it easier. In this section, you learned to use the backup and restore plugin along with automated tasks to keep GeoServer infrastructure backed up, synchronized, and stable. In the next section, we'll go over a final checklist for you to review as you put GeoServer into production.

Production checklist

In the previous section, you learned how to use automated tasks to keep your GeoServer stack backed up, in sync, and stable. In this section, we'll review a final checklist for putting your app into production. The checklist that we'll review is based on what you've learned in this book. First, we'll cover a common reference updates and authentication-related configuration. Next, we'll go over communication methods with the help of ports and protocols that must be locked down to only permit appropriate access. We'll review automated maintenance tasks. Finally, we'll revisit tuning the server for performance.

References and credentials

Follow the following points:

- Make sure all references to localhost are changed to the actual hostname when you move your work into production.
- Verify that environment variables match the node on which GeoServer is being deployed.
- Require authentication for sensitive data access to resources.
- Only accept credentials sent via secure methods such as HTTPS.
- Change the admin user password from the default.
- Create a user other than admin to do rights from OGC services and another for regular scripted REST interaction. These accounts should have privileges limited to the actions they are expected to perform.
- Create a user other than the default Postgres on Postgres to allow reading or writing to the database. Give this user limited permissions related only to the GeoServer databases which it is expected to be accessing.

Locking down

To prevent injection running of unauthorized code from the frontend, you'll need to do the following:

- Do not allow unfiltered text to be submitted through forums or requests.
- Verify that a firewall on the node is running, and that only necessary ports are active. For example, you may allow `8080` on the development firewall, but you may be only allowing `443` on the production firewall for HTTPS.
- To prevent cross-site scripting attacks, do not enable course, which is turned off by default in Jetty and Tomcat through `config` files, or turn it back off if you enable it during development and testing.
- If you need to run a different machine that requires cores, make certain to limit the domain names from which score is permitted; requests are sent with the `config` file.
- If possible, use an SQL server residing on the GeoServer node, and only allow local connections, which is the default. If remote connections are necessary, be sure to limit these through the `pg_hba.conf` file. For example, on Postgres, only allow SSL TLS connections from the address where the GeoServer node resides. This will need to be a stable address or a range.

- Definitely use HTTPS whenever sensitive information needs to be sent from a remote browser to the server. Do not allow this information to be received by HTTP.
- If you do not wish to go through the complexity of requesting certificates, you can also disallow any such access and require these actions to be done through remote access, which uses a secure tunnel in the case of Remote Desktop Protocol or SSH. If you are using HTTPS, you will need to sign up with a trusted authority to create certificates.

Regular maintenance

Follow the following list:

- Use automated tasks to regularly restart system components. Do archival and backup tasks and other regular resource-consuming tasks such as GWC on low-use hours.
- When possible, in combination with clustering for high availability, enable logging and monitoring, making sure these processes are limited to only what's needed, for example, using the production logging profile and doing audit monitoring to a file rather than a database.
- Use an external data directory and make sure that this is also backed up in an automated routine fashion, and do the same with database backups.

Tuning

Follow the following list:

- Use Java options in the startup scripts that are close to the expected need for your server. Provisioning too many resources can lead to crashes, while not provisioning enough can lead to poor performance.
- Use the best mix of JVM, native JI, and native image I/O libraries for your platform.
- Use tile caching to pre-render map images ahead of browser requests.
- Provision additional resources for your GeoServer node, or use clustering if monitoring logs or other benchmarking indicates problems.

- Use the control flow extension to prevent the system from being bogged down by a few users or tasks.
- Use the best data formats for your use case and use format-based methods to improve performance, such as internal overviews, indexes, and tiling.

Summary

In this chapter, you explored options for hosting your GeoServer instance and app, from physical on-premises servers to containerized virtualization in the cloud. You learned to monitor and use logs with the GeoServer stack. You discovered the benefits of automatic maintenance tasks and plugin, and backup and recovery with the GeoServer plugin, and finally, you reviewed a checklist for everything to watch out for as you put your GeoServer-based app into production.

After reading this book, you can now use WPS for publishing dynamic geospatial data processing services, leverage tile caches to pre-render map images on the server for faster map apps, use the best data formats and features for server performance, enable secure authentication using GeoServer's features and secure messaging over HTTPS, and put your app stack into production on the best on-premises, cloud-based, or container-based VM for your use case.

I want to congratulate you for learning the best practices for putting GeoServer into production. Thank you for taking this journey with me, and farewell!

Other Books You May Enjoy

If you enjoyed this book, you may be interested in these other books by Packt:

PostGIS Cookbook - Second Edition
Mayra Zurbaran et al.

ISBN: 978-1-78829-932-9

- Import and export geographic data from the PostGIS database using the available tools
- Structure spatial data using the functionality provided by a combination of PostgreSQL and PostGIS
- Work with a set of PostGIS functions to perform basic and advanced vector analyses
  Connect PostGIS with Python
- Learn to use programming frameworks around PostGIS
- Maintain, optimize, and fine-tune spatial data for long-term viability
- Explore the 3D capabilities of PostGIS, including LiDAR point clouds and point clouds derived from Structure from Motion (SfM) techniques
- Distribute 3D models through the Web using the X3D standard
- Use PostGIS to develop powerful GIS web applications using Open Geospatial Consortium web standards
- Master PostGIS Raster

Practical GIS
Gábor Farkas

ISBN: 978-1-78712-332-8

- Collect GIS data for your needs
- Store the data in a PostGIS database
- Exploit the data using the power of the GIS queries
- Analyze the data with basic and more advanced GIS tools
- Publish your data and share it with others
- Build a web map with your published data

Leave a review - let other readers know what you think

Please share your thoughts on this book with others by leaving a review on the site that you bought it from. If you purchased the book from Amazon, please leave us an honest review on this book's Amazon page. This is vital so that other potential readers can see and use your unbiased opinion to make purchasing decisions, we can understand what our customers think about our products, and our authors can see your feedback on the title that they have worked with Packt to create. It will only take a few minutes of your time, but is valuable to other potential customers, our authors, and Packt. Thank you!

Index

B

backup
  using 111, 112, 113, 114, 115
bottlenecks
  resolving 55, 58, 60, 61

C

certificates 84, 86, 88, 90, 92
clustered deployment 74, 76, 77, 78

G

GeoServer authentication 92, 94, 95, 96, 97
GeoServer
  app, hosting 103, 104, 106
  instance, hosting 103, 104, 106
  instance, monitoring 107, 108, 110
  tile caching, configuring 31, 32, 33, 35, 37
  URL 6, 58

H

HTTPS
  with TLS 83, 84, 86, 88, 92

J

Java Topology Suite (JTS) 11

O

Open Geospatial
  URL 7
OpenLayers (OL)
  secure login 97, 99, 101
OpenLayers integration 21, 22, 24, 25
OpenStreetMap
  URL 28

P

Postgres clustering
  URL 75
process chaining 14, 15, 17, 18, 20, 21
production checklist
  about 115
  maintenance 117
  references and credentials 116
  tuning 117
  unauthorized code injection execution,
    preventing 116
proxy environment
  configuring 79, 80, 82, 83

R

raster data stores
  optimizing 66, 67, 70, 72
  WCS 72, 74
recovery 111, 112, 113, 114, 115

T

tile cache
  using 43, 44, 46, 47, 48, 49, 50
tile caching
  about 27, 28, 29, 30, 31
  configuring, in GeoServer 31, 32, 33, 35, 37
  problems, resolving 50, 51, 52, 53
tile-backed OL app
  creating 37, 38, 40
tile-based coordinates
  URL 29

V

vector data stores
  optimizing 61, 63
  vector data formats 64, 66

W

WCS 72, 74

Web Processing Service (WPS)

about 5

installing 5, 8, 10

request builder 11, 12, 14

www.ingramcontent.com/pod-product-compliance
Lightning Source LLC
Chambersburg PA
CBHW080537060326
40690CB00022B/5156